Automotive Cyber Security

Shiho Kim · Rakesh Shrestha

Automotive Cyber Security

Introduction, Challenges, and Standardization

 Springer

Shiho Kim
School of Integrated Technology (SIT)
College of Engineering
Yonsei Institute of Convergence
Technology (YICT), Yonsei University
Incheon, Korea (Republic of)

Rakesh Shrestha
School of Integrated Technology (SIT)
College of Engineering
Yonsei Institute of Convergence
Technology (YICT), Yonsei University
Incheon, Korea (Republic of)

ISBN 978-981-15-8052-9   ISBN 978-981-15-8053-6 (eBook)
https://doi.org/10.1007/978-981-15-8053-6

© Springer Nature Singapore Pte Ltd. 2020
This work is subject to copyright. All rights are solely and exclusively licensed by the Publisher, whether the whole or part of the material is concerned, specifically the rights of translation, reprinting, reuse of illustrations, recitation, broadcasting, reproduction on microfilms or in any other physical way, and transmission or information storage and retrieval, electronic adaptation, computer software, or by similar or dissimilar methodology now known or hereafter developed.
The use of general descriptive names, registered names, trademarks, service marks, etc. in this publication does not imply, even in the absence of a specific statement, that such names are exempt from the relevant protective laws and regulations and therefore free for general use.
The publisher, the authors and the editors are safe to assume that the advice and information in this book are believed to be true and accurate at the date of publication. Neither the publisher nor the authors or the editors give a warranty, expressed or implied, with respect to the material contained herein or for any errors or omissions that may have been made. The publisher remains neutral with regard to jurisdictional claims in published maps and institutional affiliations.

This Springer imprint is published by the registered company Springer Nature Singapore Pte Ltd.
The registered company address is: 152 Beach Road, #21-01/04 Gateway East, Singapore 189721, Singapore

Preface

The goal of this book is to provide a detailed, in-depth, state-of-the-art description of vehicle connectivity and cybersecurity with respect to developments, technologies, inventions, and services. This book outlines the development of functional safety and cyber security, threats, and creative innovations in intelligent and autonomous vehicles. The chapters of this book offer a suitable context for understanding the complexities of the connectivity and cyber security of intelligent and autonomous vehicles. A top-down strategy was adopted to introduce the vehicle features and functionality. This book thus provides important information on the cyber security challenges faced by the autonomous vehicles, and it seeks to address the mobility requirements of users with their need for desire, comfort, safety, and security. This book consists of eight chapters contributed by academia, practitioners, and researchers from reputed universities from different countries.

Chapter 1 "Introduction to Automotive Cybersecurity" provides a brief overview of the topics covered in this book. It gives a brief summary of the automotive industry that explains the technological revolution that has transformed the automotive industry and eventually moves it toward new technologies such as intelligent autonomous vehicles to smart mobility. Chapter 2 "Intelligent Autonomous Vehicle" focuses on the development of the intelligent and autonomous vehicles empathizing the history and classification of autonomous vehicle driving levels based on SAE and NHTS classification. This chapter provides the latest trends and state-of-the-art intelligent and autonomous vehicle technologies and provides brief information regarding inter-vehicle and intra-vehicle communication. It discusses the technical global status of the autonomous vehicle industry, megatrends, technology adoption battle, market demand, and automotive cyber security. Chapter 3 "Security and Privacy in Intelligent Autonomous Vehicles" gives a detailed description of cryptography and cyber security used in intelligent and autonomous vehicles. It provides information related to security and privacy threats in intelligent and autonomous vehicles. It gives an overview of in-vehicle threat modeling, connected vehicle vulnerabilities, security, and privacy issues in vehicular networks. It gives a brief introduction on trust management issues in intelligent and autonomous vehicles and discusses on how blockchain can be used as a security mechanism in intelligent vehicles. Chapter 4 "In-Vehicle Communication and Cyber Security" provides a more detailed treatment of the in-vehicle communication and cyber security issues. It discusses in-vehicle electrical and electronic systems, introduces specialized advanced driver-assistance systems (ADAS) in more detail, and provides in-depth information regarding vehicle sensors, in-vehicle network types, and in-vehicle architecture and topology. It presents in-vehicle communication system, provides the functional safety as well as vehicle cybersecurity, and discusses its issues and challenges. Chapter 5 "AUTOSAR Embedded Security in Vehicles" is an invited chapter from Mr. Adi Karahasanovic, who is an expert in threat modeling for automotive cyber security and AUTOSAR. He has written Chap. 5, which is an extended version of his paper that was presented in ESCAR Conference in 2017. This chapter discusses on how to adapt and apply the threat modeling method common to the automotive industry. The cumulative benefit is accomplished by offering two methods of threat analysis, i.e., TARA and STRIDE that are uniquely tailored to the connected vehicle definition and can be further used by automobile experts. The three libraries developed using the TARA and STRIDE method frameworks are a strong starting point for all applications in the future. The two methods are implemented successfully based on the AUTOSAR norm for the connected car and the underlying software architecture. Chapter 6 "Inter-Vehicle Communication and Cyber Security" provides in-depth knowledge on vehicular ad hoc network (VANET) technology types of inter-vehicle communication. It gives a clear definition of DSRC and cellular vehicular networks and their evolution and adoption in the autonomous vehicular environment. More specifically, this chapter introduces unique features of the cellular V2X based on 5G technology such as data control software-defined network (SDN), scalable network architecture and topology, edge cloud computing like cloud/fog computing and processing, and application-oriented design as part of smart vehicles. As the chapter progresses, it focuses on security and functional safety related to V2X communication. It discusses the cybersecurity of the intelligent and autonomous vehicles against different types of attack vulnerabilities, hacking, associated risks, and their prevention and solutions. The objective of this chapter is to promote and give information on functional safety and cyber security and shed light on the challenges and issues still that need to be addressed. Chapter 7 "Internet of Vehicles, Vehicular Social Networks, and Cybersecurity" presents the Internet of Things (IoT) in vehicular networks known as IoV. First, it provides an overview of IoV network model, IoV layered architectures, security challenges and flaws in the IoV, and then security requirements and attacks in IoV along with the applications of IoV. This chapter presents machine learning techniques in vehicular networks. A strong conceptual framework for cybersecurity of cyber-physical vehicle networks is introduced based on the principles of artificial intelligence, deep neural network (DNN), and deep learning (DL) and the value of cybersecurity with respect to various types of attack scenarios along with the applications. In addition, the features of attack taxonomies, and surfaces and vulnerabilities to automotive attacks, are described along with the description of attack surface intrusion points in vehicles and the associated risks. Chapter 8 "V2X Current Security Issues,

Standards, Challenges, Use Cases, and Future Trends" discusses the international cyber security standardization in V2X communication and V2X frequency allocations. We introduce different types of standardization, associations, and organization working in DSRC and C-ITS protocols, and its associated risk factors and security issues as well as 5G V2X testbed and its use cases. It addresses the effect of electric vehicles in intelligent and autonomous vehicle and carsharing applications. This chapter introduces carhailing and ridesharing services as a promising solution to minimizing private vehicle utilization in a community, thus minimizing the need for parking spaces, reducing traffic congestion, and contributing to emission reductions. It includes companies providing taxi services/carhailing and ridehailing electronic transportation networks that provide ride services through their respective smartphone apps. In the end, this chapter provides insight on future autonomous vehicle trends, challenges, and cyber-attacks.

This book offers a structure through which the relevant requirements can be analyzed and adopted by the reader. The aim of this book is to address a wide range of topics of vehicle connectivity and cyber security and to deliver a synopsis to consider several related issues to these topics. We hope the reader of this book will take advantage by its diverse coverage of topics on automotive cybersecurity. The goal of this book is to encourage autonomous vehicle specialists as well as students, who have an interest in autonomous vehicles, share information, and provide greater transparency.

Incheon, Korea (Republic of) Rakesh Shrestha
  Prof. Shiho Kim

Acknowledgements

We would like to express our sincere gratitude to all those who devotedly helped in completing this book. We are thankful to Dr. Madhusudan Singh for his initial effort in planning, vision, and draft during the starting phase of the chapters. We would like to thank Dr. Singh for contributing sections "Cyber Security in Automotive Technology" and "Vehicular Ransomware Attack" in Chap. 1. We would like to thank Mr. Adi Karahasanovic for accepting our invitation and his contribution in writing Chap. 5. His chapter on "AUTOSAR Embedded Security in Vehicles" provides in-depth information on automotive cyber security and threat modeling. We are thankful to Dr. Rojeena Bajracharya for her contribution in writing sections such as "Machine Learning in Vehicular Networks" and "Vehicular Social Network" in Chap. 7, and her additional contributions in sections "Competition over V2X Technology Adoption" and "Standardization for V2X Communication and Frequency Allocation" in Chap. 8. We would like to thank Dr. Bajracharya for her effort in reviewing several chapters as well as drawing many of the figures and comparison tables in this book. I am also thankful to Prof. Seung Yeob Nam for his useful suggestions and expertise on cyber security in vehicular communication networks that helped to enhance the quality of Chap. 6.

We gratefully acknowledge the following organization, industry, research centers, and sponsors for their generous financial and thoughtful contribution, while writing and completing this book. We appreciate the tremendous support from Ministry of Science and ICT (MSIT), Korea, under the "ICT Consilience Creative Program" (IITP-2019-2017-0-01015) supervised by the IITP (Institute for Information & Communications Technology Planning & Evaluation), and Seoul metropolitan city through the project for writing the chapters. We are also thankful for providing support and incentive to complete this book, which is a part of research activities of SKT-Yonsei Cooperative Autonomous Driving Research Center supported by the SK Telecom's ICT R&D Center.

Finally, we are thankful to all our family members and specially to my mother, colleagues, and all individuals who helped directly or indirectly, and without them, it would not have been possible to complete this book in a timely manner.

Incheon, Korea (Republic of)   Rakesh Shrestha
                                                                                                                        Prof. Shiho Kim

Contents

1	**Introduction to Automotive Cybersecurity**	1
	1.1 Overview	1
	1.2 Introduction	2
	1.2.1 Security and Its Impact	2
	1.3 Cyber Security in Automotive Technology	4
	1.3.1 The Rising Threat	5
	1.4 Vehicular Ransomware Attack	6
	1.4.1 Vehicle Ransomware Attack Scheme	6
	1.5 Overview of Topics	7
	References	13
2	**Intelligent Autonomous Vehicle**	15
	2.1 Overview	15
	2.2 History of Intelligent and Autonomous Vehicle	15
	2.3 Classification of Autonomous Vehicle Based on Driving Levels	17
	2.3.1 SAE and NHTS Classification	18
	2.4 State of the Art of Intelligent and Autonomous Vehicle Technologies	19
	2.4.1 Autonomous Vehicle	20
	2.4.2 Connected Vehicle Technology	23
	2.5 Battle for Adoption	30
	2.6 Market Demand of Automotive Cyber Security	30
	2.7 Summary	31
	References	32
3	**Security and Privacy in Intelligent Autonomous Vehicles**	35
	3.1 Overview	35
	3.2 Cryptography Introduction	35
	3.3 Cryptography Objective	36

		3.3.1	Confidentiality	37
		3.3.2	Data Integrity	37
		3.3.3	Authentication	37
		3.3.4	Non-repudiation	37
	3.4	Cryptographic Primitives		38
		3.4.1	Symmetric Key or Secret Key Encryption	38
		3.4.2	Asymmetric Key or Public Key Encryption	40
		3.4.3	Digital Signatures	41
		3.4.4	Homomorphic Encryption	42
	3.5	Cyber Security in Intelligent and Autonomous Vehicles		44
		3.5.1	Cyber Security Framework	45
		3.5.2	Cybersecurity Layers by Design	47
		3.5.3	Threat Modeling Method (TMM)	48
		3.5.4	HARA and TARA Safety and Security Methods	52
		3.5.5	Security and Privacy Threats in Vehicular Networks	54
		3.5.6	Autonomous Vehicle Cyber Security	54
		3.5.7	Connected Vehicle Security	57
		3.5.8	Trust Management in VANET	61
		3.5.9	Blockchain as a Security in VANET	62
	3.6	Summary		64
	References			64
4	**In-Vehicle Communication and Cyber Security**			**67**
	4.1	Overview		67
	4.2	In-Vehicle System		67
		4.2.1	Vehicle Electrical and Electronic System	68
	4.3	In-Vehicle Communication		73
		4.3.1	In-Vehicle Sensing Technologies	74
		4.3.2	In-Vehicle Network (IVN) Systems	75
	4.4	In-Vehicle Network Architecture and Topology		79
	4.5	Functional Safety and Cybersecurity		82
	4.6	In-Vehicle Cybersecurity Issues and Challenges		83
		4.6.1	Challenges of IVN Architecture	83
		4.6.2	In-Vehicle Onboard Ports, Threats, and Countermeasures	85
	4.7	Cyber Security in In-Vehicle Network (IVN)		87
		4.7.1	In-Vehicle Network (IVN) Security Threats	87
		4.7.2	Cybersecurity Protection Layers	89
		4.7.3	Cybersecurity for ECU	93
	4.8	Summary		95
	References			95

5 AUTOSAR Embedded Security in Vehicles ... 97
- 5.1 Overview ... 97
- 5.2 Introduction ... 98
  - 5.2.1 Background ... 99
- 5.3 Threat Models for the Automotive Domain ... 107
  - 5.3.1 Adaptation of TARA ... 107
  - 5.3.2 Adaptation of STRIDE ... 108
- 5.4 Applying the Adapted Threat Models to the Automotive Domain ... 109
  - 5.4.1 TARA ... 109
  - 5.4.2 STRIDE ... 113
- 5.5 Results ... 114
  - 5.5.1 TARA ... 114
  - 5.5.2 STRIDE ... 115
  - 5.5.3 Related Work ... 116
  - 5.5.4 Discussion and Future Work ... 117
- 5.6 Conclusion ... 118
- References ... 118

6 Inter-Vehicle Communication and Cyber Security ... 121
- 6.1 Overview ... 121
- 6.2 Connected Vehicles ... 122
  - 6.2.1 VANET Technology Overview ... 122
  - 6.2.2 Types of Communications Technology in Connected Vehicle ... 123
- 6.3 State-of-the-Art Technologies in VANET ... 125
  - 6.3.1 DSRC-Based V2X ... 125
  - 6.3.2 Cellular-Based V2X ... 125
  - 6.3.3 Hybrid V2X Technology ... 130
  - 6.3.4 C-V2X Applications and Requirements ... 131
- 6.4 Role of Edge Computing and SDN in V2X ... 131
- 6.5 Connected Vehicle Cyber Security ... 134
  - 6.5.1 WAVE Communication Cybersecurity ... 134
  - 6.5.2 Security and Privacy in V2X Communication ... 136
- 6.6 Trust Management in V2X Communication ... 137
- 6.7 Homomorphic Encryption in VANET ... 140
- 6.8 Blockchain in V2X Communication ... 142
- 6.9 Safety Standards for IAV ... 144
- 6.10 Summary ... 145
- References ... 145

7 Internet of Vehicles, Vehicular Social Networks, and Cybersecurity 149
- 7.1 Overview 149
- 7.2 Internet of Vehicles (IoV) 150
  - 7.2.1 IoV Network Model 152
  - 7.2.2 IoV Layered Architecture 154
  - 7.2.3 Security in IoV 157
  - 7.2.4 IoV Security Requirements and Attacks 157
  - 7.2.5 Challenges in IoV 159
  - 7.2.6 IoV Applications 161
- 7.3 Machine Learning in Vehicular Networks 163
  - 7.3.1 Types of Machine Learning Techniques 164
  - 7.3.2 Type of ML in Vehicular Networks 165
  - 7.3.3 Cybersecurity Solutions Based on ML in Vehicular Networks 167
  - 7.3.4 Attacks on Machine Learning/Deep Learning 170
  - 7.3.5 Application of Machine Learning in Vehicular Networks 170
- 7.4 Vehicular Social Network 175
  - 7.4.1 Applications of VSN 178
  - 7.4.2 Security Issues 179
  - 7.4.3 Privacy Issues 180
- 7.5 Summary 180
- References 181

8 V2X Current Security Issues, Standards, Challenges, Use Cases, and Future Trends 183
- 8.1 Overview 183
- 8.2 Standards, Regulations, and Legal Issues 184
  - 8.2.1 International Cybersecurity Standardization in Automotive Industry 184
  - 8.2.2 Standardization for V2X Communication and Frequency Allocation 188
  - 8.2.3 ITS Spectrum Recommendation and Regulation Consideration 194
  - 8.2.4 Cyber Security Standardization in V2X 194
- 8.3 Competition Over V2X Technology Adoption 195
  - 8.3.1 Challenges for DSRC V2X and Cellular V2X 198
- 8.4 V2X Use Cases 200
  - 8.4.1 Smart Mobility 200
  - 8.4.2 V2X Testbed 203

	8.5	Current Trends and Future of Intelligent and Autonomous Vehicles	204
		8.5.1 Trends in Intelligent and Autonomous Vehicles	205
		8.5.2 Autonomous Electric Vehicle and Challenges	207
		8.5.3 Cyber-Attacks in Future Autonomous Vehicles	208
		8.5.4 Challenges in Future Autonomous Vehicles	210
		8.5.5 Intelligent Autonomous Vehicle Improves Environment	211
	8.6	Summary	211
	References		212

Appendix 213

About the Authors

Prof. Shiho Kim received his Ph.D. degree in Department of Electrical Engineering from KAIST, Korea. He has a long career in semiconductor research area especially working in LG Electronics, Sandisk and LG Semicon as a Research Engineer or Semiconductor Device Engineer from 1988 to 1996. From 1997, he began to perform research in universities, Wonkwang University and Chungbuk National University. And since 2011, he has been Professor of the School of Integrated Technology at Yonsei University. He was also a Chair of IEEE Solid-State Circuit Society Seoul Chapter and Chair of IEIE Vehicle Electronics Research Group.

Dr. Rakesh Shrestha (M'19–SM'20) received his B.E in Electronics and Communication Engineering from Tribhuwan University (TU), Nepal in 2006. He received his M.E in Information and Communication Engineering from Chosun University, in 2010 and Ph.D. degree in Information and Communication Engineering from Yeungnam University in 2018 respectively. From Feb. 2010 to May 2011, he worked for Honeywell Security system as a security Engineer. From Jun. 2010 to Sep. 2012, he worked as a Core Network Engineer at Huawei Technologies Co. Ltd, Nepal. From 2018 Feb to 2019 Feb, he worked as a Postdoctoral Researcher at Department of Information and Communication Engineering, Yeungnam University, Korea. He is currently working as a Postdoctoral Researcher in Yonsei Institute of Convergence Technology (YICT), Yonsei University. He was invited as a keynote speaker in the Third IEEE ICCCS conference. He has worked as a reviewer in several renowned journal and conferences. He is currently an IEEE member and his main research interests include wireless communications, Mobile ad-hoc networks, Vehicular ad-hoc network, blockchain, IoT, homomorphic encryption, deep learning, wireless security, etc.

Chapter 1
Introduction to Automotive Cybersecurity

1.1 Overview

This chapter gives a brief description of the main topics of the book in automotive cybersecurity. The automotive industry, which comprises many companies and organizations, is one of the leading industries in the world as it is more aware of its environment and responds to it. The history of intelligent autonomous vehicles has improved more than two decades before. Ever since, the automotive industry has made a big transformation to ensure efficiency and safety by eliminating traffic accidents for both the drivers and passengers. The technological progress made in sensor and navigation systems, the Internet of Things, and different types of machine learning techniques would promote new, innovative, and accessible mobility that gives rise to intelligent and autonomous vehicles. However, security plays a vital role in intelligent and autonomous vehicles. In this book, we discuss in detail about the different levels of autonomous vehicles from Level 0 to Level 4, different types of cybersecurity issues in autonomous vehicles, and future trends and challenges in autonomous vehicles. Security must be thought as an important aspect during designing and implementation of the autonomous vehicles to prevent from numerous security threats and attacks. The purpose of this chapter is to provide a comprehensive overview of automotive connectivity and to provide a framework for discussion of the several challenges and issues related to automotive connectivity, security from a technical perspective.

We will begin with basic cybersecurity terms, its impact in autonomous vehicles, security goals, threats, and challenges. We show how a ransomware attack can affect the cybersecurity of the vehicle. We then provide a detailed overview of the autonomous and intelligent vehicle communication, i.e., in-vehicle and inter-vehicle communications and the related cybersecurity issues.

1.2 Introduction

The term security refers to the protection of critical assets of the system from malicious threats and mitigation of their impact on the system. The assets can be any valuable object or entity or an organization. These assets can have certain vulnerability, which can be exploited by malicious users or attackers for their own benefits. The malicious threats are generated by malicious users or intruders, which exploits the system's vulnerability, and access or modify the critical assets of the system. These malicious users can be an individual person or group of people or software, whose aim is to find a vulnerability or weak points in the system and attack at that point to collapse, harm, or just to gain access to the system. The security assets are the resources of the system that need to be protected against malicious threats and attacks. The security assets with its environment refer to the security context. If the security environment responds friendly, then it adds for security assets and vice versa.

1.2.1 Security and Its Impact

The automotive industry is experiencing massive changes as well as gaining huge opportunities. The automotive industry is dealing with new technologies and autonomous vehicle concepts that have the potential to turn the vehicle itself into the autonomous computerized moving device. There can be various cyber security mechanisms to provide security, and each security mechanism can have single or multiple impacts on the assets to be protected. A security mechanism can level-up the security for various assets or for all assets in the system's environment context. The security mechanism may be powerful for a limited period of time or impact in such a way that there is a trade-off among assets. The cybersecurity ensures safe access to hardware and protects damage to data during network access data injection and code injection. The cybersecurity is growing as the communication level is advancing extending from big computers to very small devices being a part of the Internet of Things (IoT) [1]. Some security mechanism can have no impact on the assets or may be some time negatively affects the system's assets. Some of the terminology used in security fields includes security objective, security mechanism, threats, vulnerabilities and attacks, defense, risk, policy, assurance, resilience, and countermeasure. The security goal is to provide safety measures to achieve the confidentiality, integrity, and availability (CIA) triad for protection of the overall system along with its peripherals. The triad CIA is as follows:

- Confidentiality: The aim of confidentiality is to protect the critical information from unauthorized users. Confidentiality for network security ensures that the critical assets are accessible only to authorize users.
- Integrity: This ensures that unauthorized users do not modify or manipulate the data or information during their network transmission.

- Availability: The availability is the last component of the CIA triad that represents the real availability of our information. Authentication methods, channel access, and systems all have to function efficiently to prevent the data and make sure that it is available when required. In short, the availability aims to ensure that data and network resources are available when requested by the authorized users.

Besides the CIA triad, authentication, authorization, and accountability (AAA) also play an important role for controlling the access to the system resources, policy enforcement, auditing, etc. The AAA is a term for controlling the access to the system resources, auditing usage, enforcing policies, and offering the details need to charge for services.

- Authentication: In general, authentication is about personal identification information. It contains the incoming request validation mechanism against certain identification credentials.
- Authorization: Authorization means that the user has the approval or privilege to perform a specific operation. Typically, the authorization process takes place after the successful authentication.
- Accountability: Accountability is the third component of the AAA framework. It provides administrators the ability to monitor the activities that users have conducted in a given situation. It is a primary way of evaluating what services have been used and how much resources users have consumed. Accountability is generally enforced by conducting audits, as well as setting up systems for making and maintaining audit trails.

- **Vulnerabilities and Attacks**

The wireless networks are more vulnerable and prone to different types of attacks as compared to the wired networks because any attacker can easily connect to an unsecure switch port without establishing any physical connection to the device. Unauthorized access is the most common vulnerability in wired and wireless networks. There is a tremendous amount of vulnerabilities in the network, and the transmission data is enormously vulnerable to attacks. Any network is attacked first by acquiring the communication channel, then obtaining data and then using the data for some malicious purposes. The network security targets for securing not only the end devices, but also the entire network and end-to-end connectivity. Other vulnerabilities that can take place after unauthorized access are as follows:

- Sniffing the data packet during its transfer to capture critical information.
- The network channel is overloaded with unauthentic data in order to subject denial of service to authentic users on a network.
- The MAC addresses of authentic hosts are spoofed to capture data and induce man-in-the-middle attack.

1.3 Cyber Security in Automotive Technology

Every year, there are huge developments of novel automotive applications and services that are leading the production in terms of cost and technology. More than 90% of automotive inventions lead to innovations in vehicle hardware and software. Hardware in the vehicles has control system, which directs the vehicles to perform various tasks on the roads while driving. Some basic tasks are listed below:

- Primary systems of the vehicles, e.g., the engine, driver assistance, driveline, electric system, brake system, dashboard, etc.
- Secondary vehicles systems, e.g., ignition, indicators, window control, wipers, lights
- Infotainment applications, e.g., navigation systems, telematics, rear seat entertainment, music and video entertainment, and GPS-based services.

Modern advancements in electrical and electronics have dramatically changed the automotive industry. These vehicles are no more completely mechanical systems after the intervention of electronics. In addition, these electronic devices have added a set of unimaginable features, improving the overall performance of the vehicles.

Nowadays, vehicles are faster, more sophisticated, new functionalities, and more efficient. These advancements result from dozens of electronic control units (ECUs) and vast communication network interconnecting them and enabling a whole new driving experience: from vehicles that can be remotely locked and unlocked, to vehicles that can be driven without a key in the ignition and can even drive or park themselves. This new driving experience is achieved by using hundreds of megabytes of code contained in the vehicle's ECUs. One can find Google's driverless vehicles driving themselves in Nevada, and they are also allowed in Florida and California. So it is just a matter of time until we will see more autonomous and smart vehicles all around the globe, probably raising driving safety; however, what about the security aspects?

Because of the current advancement in technologies, it is possible to deploy them to create numerous safety devices and protocols for the vehicles such as automatic emergency braking, forward collision warning, vehicle-to-vehicle communications, and soon fully automated vehicles. Considering that these innovations have great potential in them, vehicle manufacturers and transport authorities are attempting to improvise secure tools to protect these technologies from new challenges, relating primarily to cyberattacks.

As the vehicular systems become more advanced, the possibility of cyber attacks on these systems is increasing alongside. There is a need of applying cybersecurity principles at various components of the vehicles to govern safety, security, and protection from any kind of malicious attacks, damage, illegitimate access, or something, from cyber attacks, which may compromise the safety of the vehicle or the driver. The timeline of automotive cybersecurity history is given in Fig. 1.1 [2].

1.3 Cyber Security in Automotive Technology

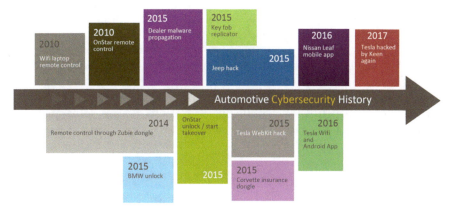

Fig. 1.1 Automotive cybersecurity history timeline

1.3.1 The Rising Threat

As mentioned above, as vehicles are getting more advanced, they use computers to control decisive operations such as brakes, stability control, and airbags functionality. Vehicles become safer on one hand, but on the other hand, the safety mechanisms are controlled by the ECUs, adding more complexity and potential attack vectors to the system. These ECUs are interconnected through a CAN bus in an unsecured way. The notion of merely securing ECUs in individual systems is inadequate in the connected and autonomous vehicle situation. There are several forms of networking such as Ethernet, cellular, Internet, Wi-Fi, Bluetooth, and V2X connected in the vehicle and linking to various networks like controller area network (CAN) bus, Ethernet that are running applications on peripherals (USB, display, sensor, LiDAR, etc.).

CAN bus is an old protocol that was introduced in the 1986, which was not designed according to security guidelines; the messages which are being sent contain an ID, their length, and the payload itself, thus message receiver does not know who the sender is and whether the message is legal. The priorities of the messages are also inferred from the message IDs, thus allowing any component connected to the bus, the ability to send high priority messages, and flooding the bus and other components.

Currently, cybersecurity problems are getting even worse. The vehicles may be remotely hacked or immobilized. Vehicles can be compromised through the wireless sensors in the vehicles, and person can be hurt physically if someone gets hold of the CAN bus and stops the vehicle suddenly through one of the external interfaces of the vehicle. Even luxury vehicles such as the Jaguar suffered from flaws, causing "blue screen of death" immobilizing them.

The weakness in several vehicle assessments is seen as extremely vulnerable security architecture with a significant number of powerful computers with wireless access such as GPS, NFC, Bluetooth, cellular, IR, Wi-Fi, etc. Taking control over such a wireless system can lead to full control over messages, which are being sent over the bus, thus entirely compromising the vehicle.

A surge of cyber attacks has recently been on the rise and has been targeting the vehicles all over the world. There is no reason to think that these machines are not part of the game because we know that they are the computers we are driving.

One might wonder why anybody would hack a car. The motive may vary, from the theft of vehicles and theft of personal information to extortion, harming the reputation of companies or even homicide or terrorism.

1.4 Vehicular Ransomware Attack

This section describes the procedure for performing a vehicular ransomware attack based on [3].

1.4.1 Vehicle Ransomware Attack Scheme

A representative ransomware attack technique is explained below as shown in Fig. 1.2.

The cyber attacker represented by (3) uses the ransom control software (5, i.e., "bot master"), distributes (c) the ransom malware (2) to their target extortion vehicles (11), behind an anonymous botnet (4), applying TOR technology as an example. The cyber attacker could then attempt (b) to insert the ransomware explicitly (c2) or implicitly (c1) over any wireless (6a) or wired network (6b) that could enter the target vehicles. When the malware enters a potential target vehicle, it utilizes the primary

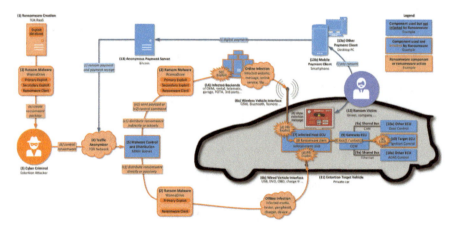

Fig. 1.2 Ransomeware attack scheme in vehicles Adapted from [2]

1.4 Vehicular Ransomware Attack

security exploit (d) of the integrated vehicle to install and deploy the ransomware client (8) on a central, well-connected in-vehicle device (7), like the infotainment unit, misusing it as a host for its further attack. From here, the ransomware client may either initially create an online connectivity back to the attacker to receive additional data ("payload") and/or more commands (e) or directly communicate (f) to a crucial target ECU like ignition control (10) via in-vehicle bus systems (9a) to execute the real onboard locking operation (g) to extortion payment from the ransom victim (12). Then, the ransomware will show the extortion message (h) and demand payment of the associated ransom. When the victim has paid the ransom (i) using an anonymous payment system (13) such as Bitcoin, the attacker would again contact his ransomware using his anonymous botnet (e2) to perform the unlocking command (f) required to free the vehicle (if feasible).

It shows that ransomware for vehicles can be developed and injected, indicating that this ransomware attack on vehicles is real and serious. As the cars are more interconnected and as digital technology begins to provide even more essential vehicle technologies and as vehicle digital technology becomes more standardized, the attack surfaces for ransomware will grow, and the value that ransomware may take hostage will increase. The community that can attack a single type of ransomware will increase.

1.5 Overview of Topics

The number of connected vehicles is growing rapidly with escalating computing and storage resources being provided for vehicles worldwide. The digital transformation of the vehicles combined with various other technologies results in connected, intelligent, and autonomous vehicles also known as self-driving vehicles and automotive cybersecurity. This book provides a comprehensive overview of automotive cyber security and framework for several issues related to connected, intelligent, and autonomous vehicles. The chapters in this book give the reader essential knowledge related to vehicular communication technologies, cyber security, vehicle embedded system securities, standards, challenges, autonomous vehicle use cases, and future trends. The chapters are as follows:

1. Introduction to Automotive Cybersecurity
2. Intelligent and Autonomous Vehicle
3. Cybersecurity and Privacy in Intelligent Autonomous Vehicles
4. In-vehicle Communication and Cyber Security
5. AUTOSAR Embedded Security in Vehicles
6. Inter-vehicle Communication and Cyber Security
7. Internet of Vehicles, Vehicular Social Networks, and Cyber Security
8. V2X Current Security Issues, Standards, Challenges, Use cases, and Future Trends

We will briefly introduce the contents of each chapter hereafter.

1.5.1. In (Chap. 2), the history and evolution of intelligent and autonomous vehicle outlook are provided. It provides a detailed overview of the of intelligent and autonomous vehicle development timeline since 1920s. It provides the classification of autonomous vehicles based on vehicle driving levels, associations between connectivity and autonomy for vehicle performance, and the state-of-the art applications. The intelligent and autonomous vehicle is the next-generation self-driving vehicle equipped with different types of advanced sensors, actuators, controllers incorporated with intelligence (e.g., machine learning techniques), and cooperative driving capability to guarantee autonomy, safety, protection, ease, and energy efficient. In Chap. 2, the in-vehicle technologies in autonomous vehicle are categorized into three different technologies, and they are as follows.

1. Sensor Technologies: The sensor technologies used in autonomous vehicles consist of LiDAR, VLS, ultrasonic ranging sevices (URD), infrared ranging, and millimeter wave radar (MWR), etc. The LiDAR consists of GPS, scanner, and laser technology to generate 3D information about a particular area, providing remote sensing based on pulses of light.
2. Vision Technologies: The vision technologies consist of Stereo Vision System (SVS), HD cameras, Black box, or CCTVs. It helps in forensics and takes necessary actions by recording visual information with high confidentiality, authenticity, and integrity.
3. Positioning Technologies: The positioning technologies include GPS, radar cruise control, and radar-based obstacle detections (RBOD). In autonomous vehicles, the GPS receiver can be used in conjunction with the Doppler radar speedometers, RCC, and RBOD to provide precise vehicle location and active location validation.

This chapter discusses about different types of vehicular ad hoc network (VANET) technologies based on dedicated short-range communications (DSRC) or 802.11p Protocol and cellular technologies. It provides DSRC standard suites and applications of VANETs such as safety applications, cooperative collision avoidance (CCA), emergency warning messages (EWM), traffic managements, and infotainment applications. In the end, it discusses about the market demand of automotive cybersecurity for current and future environment. It discusses the use of cellular technology to support vehicular communication known as LTE for vehicles (LTE-V) that is being researched and developed. It is an alternative technology for ITS that uses existing cellular base stations making urban transportation more manageable and efficient.

1.5.2. Chapter 3 mostly focuses on cybersecurity and privacy in intelligent and autonomous vehicles. It provides different types of security and encryption schemes that can be used in autonomous vehicles. The cyber security in intelligent and autonomous vehicles can be a combination of physical security, information security, policies, standards, legislation, and risk mitigation

1.5 Overview of Topics

strategies. It also discusses about different security elements used in intelligent and autonomous vehicles. The malicious attackers use different means of attack strategies at different levels, and it severely impacts on the vehicle's physical and cyber security system. The malicious nodes can tamper the vehicle sensors such as onboard systems, in-vehicle sensors and can intercept message exchange between the vehicles in the in-vehicle communication. The malicious nodes can be an insider attacker that may attack both the in-vehicle and inter-vehicle communication. Several attacker models have been demonstrated. It provides some of the security and privacy threats such as fake information attack, message replay attack, integrity, non-repudiation, access control, and privacy attack. The vulnerability is the weak point in the autonomous vehicle system that are misused and easily attacked by the attackers for their own advantages. A detailed autonomous vulnerability taxonomy of the vehicles is given, and solution mechanisms such as preventive, active, and passive defense are also provided. It also deals with the privacy of the autonomous vehicles. Some of the privacy measures discussed are cryptography-based schemes, trust management schemes, and blockchain schemes.

1.5.3. Chapter 4 mainly focuses on the in-vehicle communication systems and cyber security issues. In this, we concentrate on safety, cybersecurity, and privacy of the embedded automotive vehicles within the automotive domain. It discusses about in-vehicle electrical and electronic systems, introduces specialized advanced driver assistance systems (ADAS) in more details, and provides in-depth information regarding vehicle sensors, in-vehicle network types, and in-vehicle architecture and topology. We discuss seven categories of vehicle electrical and electronic (VEE) used in the in-vehicle system. The vehicles use ECUs to communicate with other control units, sharing vital vehicle information via the LAN protocol. Moreover, we discuss about the different types of IVN protocols such as CAN, FlexRay, Automobile Ethernet, LIN, and MOST along with their security threats. We present the IVN network architecture and its challenges on OBD-II ports, threats, and countermeasures. In the end, we discuss the cybersecurity in IVN and present the cybersecurity protection layer for the in-vehicle systems. It provides the functional safety as well as vehicle cybersecurity and discusses its issues and challenges. The safety and security in automotive engineering are closely related with each other, and if there is a worthy interaction between them, then they could get huge benefit from each other.

1.5.4. Chapter 5 focuses on embedded security systems in vehicles such as AUTOSAR. We discussed AUTOSAR and different threat models and risk assessment for automotive vehicles. Within this chapter, we analyze two methods of threat modeling commonly used in the computing industry and determine their eligibility to the connected vehicle. We further suggest improvements to these methods of threat modeling to make them more applicable to the fundamental architecture of applications used in today's vehicles, i.e., AUTOSAR. The first method, TARA, reflects an attacker-centered approach while the second method, STRIDE, explores the system's information infrastructure and is part of the software-centered approach. The two methods are implemented successfully based on the AUTOSAR norm for the connected car and the underlying software architecture. The TARA method was developed by security experts from Intel Security and is based on three groups of collected data, denoted as libraries, and they are Threat Agent Library (TAL), Methods and Objectives Library (MOL), and Common Exposure Library (CEL). The three libraries developed using the TARA and STRIDE method framework are a strong starting point for all applications in the future. This chapter shows the effectiveness of these approaches, including the real validation of STRIDE tests on real devices. The domain experts will be able to include them in their tool set for future research and analysis.

1.5.5. In Chap. 6, we focus on the inter-vehicle communication system and its cybersecurity issues. We describe in detail the various types of connected vehicle technology and its security issues. The cybersecurity protects the system, or networks from malicious cyberattacks that interrupt the normal communication in the network or thwart the functioning of the system or steal the sensitive information. This section discusses about the cybersecurity of the intelligent and autonomous vehicles against different types of attack vulnerabilities, hacking, associated risks, their preventions, and solutions. We discuss the different types of security and privacy issues and security requirements in connected vehicles. This section especially focuses on security and functional safety related to V2X communication.

We examine recent developments in connected vehicle technology and provide different types of security issues in connected vehicles. The autonomous vehicles based on a limited sensing range cannot be fully trusted. The connected vehicles offer a broader understanding of the surrounding environment and help vehicles make smarter decisions about the messages exchanged between the neighboring vehicles and the RSU. This helps in planning the future travel route in a safe and secure way. The vehicle-to-everything (V2X) is the main communication technology for future VANETs that helps vehicles to obtain a wide range of road information in real time that significantly improves driving safety, traffic efficiency as well as provides infotainment services. In this chapter, we overview the V2X technologies and discussed vehicle-to-infrastructure (V2I), vehicle-to-network (V2N), vehicle-to-vehicle (V2V), and vehicle-to-pedestrian (V2P) communication. This chapter also discusses, in detail, DSRC-based V2X technology to support cooperative awareness applications such as vehicle warning, emergency brake light, and vehicle platooning. However, these applications are suitable only for low density of vehicles with low bandwidth and it is not suitable for high density of vehicles and infotainment applications. So, the third-generation partnership project (3GPP) works toward modifications of radio access suitable for V2X communications known as cellular V2X or C-V2X. In C-V2X, it extends the cellular device-to-device (D2D) communications specification by introducing two more operational modes dedicated to V2V

1.5 Overview of Topics

communications, i.e., Mode 3 and Mode 4. With the advancement in C-V2X technology, the 3GPP enhanced the LTE technologies and released the 5G C-V2X, and recently the 3GPP accelerated the work in new radio technology and introduced and 5G NR C-V2X in Rel. 16, which have backward compatibility features. There are also hybrid technology that combines both the DSRC and cellular technology for vehicular communication. Later in the chapter, it provides the evolution of C-V2X toward 5G technology for autonomous vehicles, as well as its applications and requirements. In the end, we explain the security, privacy, and trust management issues in connected vehicular technologies. We discuss the trust management issues, homomorphic encryption, and blockchain as a security in V2X communication.

1.5.6. In Chap. 7, we discuss about the integration of ITS with the features of IoT, which is called the Internet of Vehicles (IoV). Over time, the IoV has developed from the conventional vehicular networks to advanced infrastructures and other equipment for the ITS and road transport. The IoV is a complex Internet-connected vehicular network where the vehicles are equipped with different types of sensors that collect data from other vehicles and road infrastructures and send them to the cloud. Some of the characteristics of IoV are complex communication, dynamic topology, high scalability, localized communication, and high processing capacity: We present the detailed IoV architecture layer, security attacks, security requirements, and challenges. Some of the applications of IoV can be seen in intelligent transportation system (ITS), business-related applications, and smart city applications. The existing IoV architecture lacks security requirements such as authentication, authorization, and trust-related issues. This chapter discusses the machine learning techniques used in IoV and presents different solution approaches to overcome various attacks based on machine learning.

We also discuss about new emerging paradigm called vehicular social network (VSN), which is the integration of social networks and IoV that builds a social relationship among the vehicles as well as the drivers of the vehicles. In addition to the social relationship, the VSN can combine vehicle communication networks with the human factors that influence vehicle communication between vehicle drivers. In VSN, each vehicle is capable of creating social links with neighboring vehicles, drivers, and smart devices on an autonomous basis. The versatility of vehicles will carry out the features of social networks in which vehicles display similar gestures and habits when driving on the motorway. The VSN architecture consists of three layers, i.e., IoVs, VSNs, and social networks. It also features different applications of VSN and at the end discusses several types of attacks challenging issues.

1.5.7. Chapter 8 provides a detailed information on international cybersecurity regulations, standardizations, and types of organization working in DSRC and C-ITS protocols. The intelligent and autonomous vehicles are at the peak of a major breakthrough in vehicle communication and safety on the road. It is going to be fully implemented soon, which will change the human mobility behavior. Thus, the international technological infrastructure around autonomous vehicle implementation is already under demand to develop a new set of standards while replacing the existing rules. There are several organizations, consortiums, associations, and authorities working toward the development of new standards, policies, and regulations.

They have been categorized into three regions, viz. Europe, USA, and global standardizations. Several cybersecurity regulations, initiatives, and projects have been carried out under Europe initiatives. Some of them are as follows: In 2006, the SEcure VEhicle COMmunication (SEVECOM) project started to deal with security of vehicular communications and inter-vehicular communications. It provided solutions to the problem that are specific to the road traffic information. In 2008, the E-Safety Vehicle Intrusion protected Application (EVITA) project started, and its goal was to design and verify OBU prototypes and provide e-safety by securing the electronic components of vehicles from tampering. From 2008 to 2012, the SimTD project was carried out in Germany. Its objective was to increase road safety and improve the traffic efficiency based on V2X communication. The result of this project can be applied in the categories like traffic and value-added service. The 7th Framework Program of the EU commission started the Open VEhiculaR SEcurE (OVERSEE) project in 2010 and ended in 2012. OVERSEE provided standard, secure, and generic communication application platform for vehicle and enhanced the efficiency and safety of the road traffic. Similarly, the framework funded another project called Preparing Secure Vehicle-to-X Communication Systems (PRESERVE) in 2011 and ended in 2015. The PRESERVE objective was to design an integrated V2X security architecture (VSA), implement the architecture, and field test the VSA system. The CAR2CAR Communication Consortium (C2C-CC) was established in Europe, which is a consortium of leading European and international vehicle manufacturers, equipment suppliers, engineering firms, road operators, and research institutions. Similarly, the AUTomotive Open System Architecture (AUTOSAR) is very popular in-vehicle software standardization organization for intelligent and autonomous vehicles. The AUTOSAR is a global consortium of automakers, suppliers, service providers, vehicle industry, semiconductors, and Software Company. In the USA, the Society of Automotive Engineers (SAE) and International Organization for Standardization (ISO) jointly worked together to develop the current state-of-the-art cybersecurity standards for vehicles in two areas, i.e., road vehicles and ITS. The SAE and ISO co-chaired and worked as a Joint Working Group (JWG) to introduce ISO/SAE 21434 under a new agreement. The Institute of Electrical and Electronics Engineers Standards Association (IEEE-SA) introduced 802.11p standard to support Wireless Access in Vehicular Environments (WAVE) for ITS applications. As for international initiatives, the International Organization for Standardization (ISO) and SAE work together for the vehicle cybersecurity standards. The ISO introduced ISO 26262, which is an international risk-based standard for functional safety of electronic and electrical systems in vehicles derived from IEC 61508. Similarly, the United Nations

1.5 Overview of Topics

Economic Commission for Europe (UNECE) is preparing a certification for a Cyber Security Management System (CSMS) that mandates the approval of the vehicles according to the requirement of the recent document proposal. It is working toward a global standardization and regulation on cybersecurity of the vehicles focusing on the vulnerability issues like Over the Air (OTA) issues. Likewise, there are several safety and cybersecurity standards and projects in the automotive industry, and the detail information and timeline are given in Sect. 2 of Chap. 8. Similarly, Chap. 8 discusses the V2X technology based on DSRC and cellular network and its adoption. It showcases the 5G V2X test bed and its use cases around the globe. Finally, it presents the future of intelligent and autonomous vehicles, cybersecurity issues, and solutions based on machine learning.

References

1. R. Shrestha, S. Kim, Integration of IoT with blockchain and homomorphic encryption: challenging issues and opportunities, in *Advances in Computers*, vol. 115, eds. by S. Kim, G. C. Deka, and P. Zhang (Elsevier, 2019), pp. 293–331
2. Y. Xiang Gu, The industrial challenges in software security and protection, in *The 9th International Summer School on Information Security and Protection*, (2018), pp. 1–138
3. M. Wolf, T. Enderle, R. Lambert, A. Schmidt-derrick, WannaDrive? Feasible attack paths and effective protection against ransomware in modern vehicles, in *15th ESCAR Europe* (2017), pp. 1–14

Chapter 2
Intelligent Autonomous Vehicle

2.1 Overview

This chapter provides an overview of the developments in intelligent and autonomous vehicle (IAV) that accompany in-vehicle and inter-vehicle connectivity. This chapter gives detailed information regarding the evolution of intelligent and autonomous vehicles, its classification based on vehicle driving levels, associations between connectivity and autonomy for vehicle performance and the state-of-the-art applications.

The intelligent and autonomous vehicle technologies can be categorized into autonomous vehicles also known as self-driving vehicles and Cooperative Intelligent Transport Systems (C-ITS) in Europe [1], which is also known as connected vehicle technologies in the USA [2]. The autonomous vehicles are based on the combination of different technology and sensors to achieve desired autonomous level. In contrast, the connected vehicle technologies or C-ITS are based on vehicular ad hoc networks also known as VANETs for transmitting beacon messages, basic safety messages, and infotainment messages in IAV. Hence, IAV is the next-generation self-driving vehicle equipped with different types of advanced sensors, actuators, controllers incorporated with intelligence (e.g., machine learning techniques) and cooperative driving capability to guarantee autonomy, safety, protection, ease, and energy-efficient.

2.2 History of Intelligent and Autonomous Vehicle

More than 130 years ago, Karl Benz introduced the first automobile or motorized vehicle, which is later known as Mercedes-Benz. Since then the automobile industries have taken a huge transformation for improving traffic efficiency, providing safety and security to the drivers as well as passengers by eliminating traffic accidents. The history of intelligent autonomous vehicles has improved more than two decades before [3]. During the early 80s, the research in intelligent transportation system (ITS) has increased significantly. Automated and intelligent vehicles have been the most significant applications of ITS. The first automated vehicles were introduced by DARPA in 1984. The DARPA funded a project called autonomous land vehicle (ALV) that demonstrated the first on-road automated vehicle by using computer vision, light detection and ranging (LiDAR), global positioning system (GPS), and robotic control technologies to direct the vehicles [4]. Before this, in 1980, a vision-guided driverless robotic vehicle was demonstrated on the road without other normal vehicles on the road by Mercedes-Benz. The vehicle had maximum speed of 63 km/hr but it could not gain any attention from automated vehicle industry [5].

The automated vehicles can perform work based on computers or machinery rather than the driver, its automatic movement and functioning is subjected to deeply dependent on artificial aids in their environment such as magnetic strips. In 1989, scientist from Carnegie Mellon University (CMU) pioneered the use of artificial intelligence (AI) on the previous AVL. They used artificial neural network on their autonomous navigation test vehicle called NAVLAB by using connected approach ALVINN (i.e., autonomous land vehicle in neural network) [6]. By using artificial intelligence in the automated vehicle, the vehicle now becomes the intelligent vehicle (IV), which has perception based on key control algorithm. During mid-90s, General Motors (GM) initiated vehicular networks by developing OnStar telematics, where vehicle and the system can communicate with each other [7]. The intelligent vehicle achieved connectivity to communicate with other vehicles and the infrastructures. The term "connected vehicle" was introduced by the US Department of Transportation (US-DoT) to indicate that the vehicles can communicate with other vehicles and infrastructures while driving on the road, but it does not mean that the vehicles will be automated. The reason being is that the wireless communication offered by the OBU informs the driver regarding hazards or collision and the driver should act to avoid the hazards. There are two types of connectivity: the first one connected by using dedicated short-range communication (DSRC) and WAVE [8], for vehicle-to-vehicle (V2V) and vehicle-to-infrastructure (V2I) communications and then extended to vehicle-to-everything (V2X) communication, which became the core of vehicular communication. The second one connected by using cellular technology (e.g., LTE and 5G) that connects vehicle to the Internet and cloud and other Internet of things (IoT). However, it is not necessary for the automated vehicles to have wireless communication functionality because automated driving systems (ADS) use different sensors to gathers driving environmental data.

During 2005 and 2007, the researchers from academic fields such as CMU, Stanford and MIT research on multi-sensor fusion and used various sensors on their test intelligent vehicles like LiDAR, GPS, inertial measurement unit (IMU), radar, and other multiple cameras. It was only in 2009, Google Inc. introduced the autonomous vehicle based on high-precision map. It is also known as self-driving car, robotic car, autonomous vehicle, driverless cars, etc. The autonomous vehicle is a computer-controlled vehicle that drives itself with the help of various sensors as mentioned above to perceive the surroundings and to identify obstacles and relevant signs. After this, there is a rapid development in the autonomous vehicles. In 2014, Elon Musk

2.2 History of Intelligent and Autonomous Vehicle

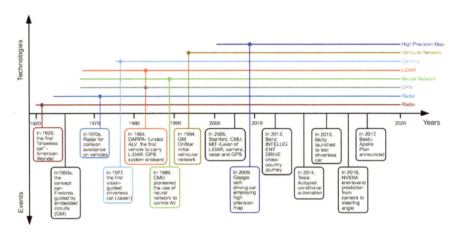

Fig. 2.1 Autonomous and intelligent vehicle development timeline

launched autonomous car called Tesla Model S with conditional automation as a Tesla Autopilot project. It has features such as auto parking at the parking lot without the driver inside the vehicle [9]. According to SAE, Tesla's automated driving features are categorized as a level 2 driver assistance system. However, in 2016 there was a fatal Tesla crash with autopilot feature in Hubei China. In the same year 2016, a GPU manufacturer, NVIDIA achieved a milestone in autonomous vehicle to foresee end-to-end prediction using camera and steering angle based on deep learning [10]. In 2017, there was a huge competition among the automotive companies and academic research fields to develop first intelligent and autonomous vehicle such as Ford, Audi, Baidu, University of Michigan, Mercedes, and many more. In 2018, Google Inc.'s, Waymo became the first self-driving technology company to launch and commercialize a fully autonomous taxi service [11]. The autonomous and intelligent vehicle development timeline is given in Fig. 2.1.

2.3 Classification of Autonomous Vehicle Based on Driving Levels

The aim of the intelligent and autonomous vehicle is to develop fully autonomous vehicles. The development of Intelligent and autonomous vehicle can be categorized into two phases: the initial phase for assisted driving and the final phase for complete self-driving without human interaction. However, it is not enough to say the vehicle is autonomous vehicle; we need to specify the level as well to avoid the confusion.

2.3.1 SAE and NHTS Classification

The Society of Automotive Engineers (SAE) issued a six-level worldwide automation standard (i.e., J3016) for IAV [12]. This six-level standard is followed by several automakers worldwide to guide their automotive manufacturing and research like Tesla, Nissan, and Toyota. In 2016, the National Highway Traffic Safety Administration (NHTSA) released a Federal Automated Vehicle Policy document mentioning the change in its original five-level of automation (i.e., from 0 to 4) to correspond with the six-level of automation produced by SAE. The NHSTA definition for autonomous level driving is very broad, so SAE developed the autonomous levels based on NHSTA levels [13]. The functionality of Level 0, Level 1, and Level 2 in SAE international is similar to NHTSA standard but there are differences in the description and functionality in the higher level as shown in Fig. 2.2. In SAE, the driver needs to take over the control of the vehicle in Levels 3 and 4 when the system requests to intervene the driving task. In SAE Level 5, the autonomous vehicle performs all the driving tasks under all conditions as done by the human driver. The only job of the driver would be to input the source and destination in the system to navigate. The full automation allows the vehicle to operate without driver. We present a brief six standard level of automation from Level 0 to Level 5, focusing on SAE standard in Table 2.1.

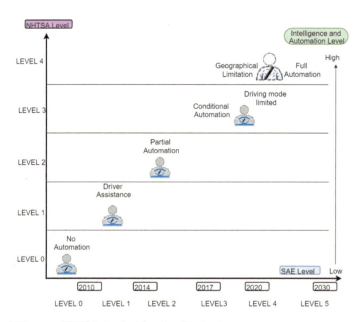

Fig. 2.2 SAE versus NHTSA standard functionality levels

Table 2.1 SAE Level standard description according to SAE [12]

SAE level	Short description	Details
0	No automation driving	Vehicle is fully controlled by the driver in all conditions
1	Driver assistance	Achieves either lateral or longitudinal vehicle motion when activated (but not both simultaneously)
2	Partial automation	Achieves both lateral or longitudinal vehicle motion when activated
3	Conditional automation	Provides partial driving task and monitor the driving environment but driver should be ready to take over the control when system requests
4	High automation	System can conduct the entire driving task and monitor the driving environment even if the driver does not respond to system's request to intervene in specific condition and geographical location
5	Full automation	System can conduct entire driving task, under all dynamic driving condition once programmed with the destination in all environments manageable by the driver

2.4 State of the Art of Intelligent and Autonomous Vehicle Technologies

With the development of intelligent and autonomous vehicle technology, it has fascinated an unexceptional level of attention from different fields like automotive industry, academic world, and media while raising speculations about the consequences and effects of the automated self-driving on the society such as traffic flows, road traffic, environmental issues, vehicle economy, road safety, privacy of the passengers, and cybersecurity. Today, the driver can drive the vehicle safely on the road with supreme comfort because of the growing reliability and performance of hardware components and software technologies that empower complex automotive functions.

There are two basic autonomous vehicle architecture from technical viewpoint, viz. autonomous vehicle and connected vehicle. The first is based on vehicle platform, i.e., autonomous vehicle that largely depends on intra-vehicle sensors or in-vehicle sensors. The varieties of sensors attached in the vehicle and multisensory fusion technology obtain perception about the environment and obstacles using signal processing, then make decisions, and control the vehicle. The second is based on connected vehicle technology largely depend on vehicular ad hoc networks (VANETs). The vehicle receives the environmental information and road information through the VANET communication technology.

We briefly discuss the autonomous vehicle and connected vehicle in IAV.

2.4.1 Autonomous Vehicle

The automated vehicles refer to those vehicles, which can be automated based on human intervention or some kind of human assistance. The automated vehicles cannot select the destinations or route for the drivers independently. On the other hand, the autonomous vehicles should be able to perform actions independently without the intervention of the driver and respond to the accidents in level five. The autonomous vehicles are based on combination of different technology and multi-sensors to achieve the desired autonomous level. Since a large number of sensors are deployed in the vehicle to achieve autonomy, those sensors need to communicate with each other efficiently to achieve autonomous driving. In the next subsection, we will discuss the in-vehicle communication system in autonomous vehicles.

2.4.1.1 In-Vehicle Communication

The in-vehicle communication refers to the intra-vehicle communication where all the internal components like telematics, sensors, and actuators communicate with each other using different communication medium like BUS system. There are different vehicle domains that have specific requirements, which led to the development of a large number of automotive BUS network such as Local Interconnect Network (LIN), Control Area Network (CAN), FlexRay, Media Oriented Systems Transport (MOST), automotive Ethernet, and Low-Voltage Differential Signaling (LVDS) AVB [14] as shown in Fig. 2.3. More will be discussed in detail in Chap. 4.

To achieve full self-driving intelligence, the vehicles need to observe their own condition, surrounding environment, and other situations beyond their visual range. The perception of vehicle self-state and decision making is based on the installed high-precision sensitivity sensors such as pressure, engine temperature, and speed sensors. The driving safety can be ensured by using the sensors installed in the vehicle that captures perception of the surrounding environment of the vehicle such as obstacles on the road, driving condition, and environmental condition. In case of high-speed driving, the safety of the vehicle can be ensured by long-range sensors or other communication modes. Based on these information obtained from sensors, the vehicle will accomplish the self-driving control independently based on environment sensing, decision making, and other driving controls. The in-vehicle technologies can be further categorized into three different technologies. They are (a) sensor technologies: Sensor technologies include LiDAR, VLS, ultrasonic tracking systems (URD), infrared, and Millimeter Wave Radar (MWR). The LiDAR consists of GPS, scanner, and laser technology for the generation of 3D information in a specific area. The VLC is a visible light-based system for transferring information from one location to another but with a small range of just about 50 m. (b) Vision technologies: It comprises of Stereo Vision System (SVS), HD cameras, Black box, or CCTVs. These provide graphic images for the study of sensitive circumstances such as traffic collisions. SVS reconstructs a 3D video using several viewpoints and

2.4 State of the Art of Intelligent and Autonomous Vehicle Technologies

Fig. 2.3 Autonomous vehicle functional components with sensor fusion

structured light source based on active and passive methods. (c) Positioning technologies: Some of the positioning technologies are GPS, radar cruise control, and radar-based obstacle detections (RBOD). In autonomous vehicles, the GPS receiver can be used in combination with the Doppler radar speedometers, RCC, and RBOD to provide precise vehicle positioning and active position confirmation.

2.4.1.2 In-Vehicle Networking Types

Since advanced vehicles such as electric motors, hybrid or self-driving vehicles need more rigorous real-time information, and to satisfy these criteria, in-vehicle networking system implementation is rapidly emerging. It is also due to the complex and heterogeneous architectures with FlexRay or Ethernet networks. Since a large number of sensors are used in automotive vehicles, a big amount of data is produced every second. And all those sensors need to communicate with each other with strict minimum latency. Different components in the vehicles require different bandwidth and latency and continue to increase on complexity along with the increase in number of electronic control unit (ECU) used in automotive vehicles. The automotive Ethernet offers a high-speed communication for in-vehicle communication. The IEEE 802.3bw standardized a gigabyte solution of automotive Ethernet in 100BASE-T1 [15]. The in-vehicle networking system is given in Fig. 2.4.

1. FlexRay: A common automotive network called FlexRay was created by the alliance of major companies from the automotive industry and leading suppliers.

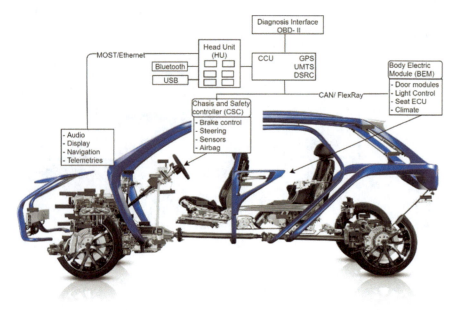

Fig. 2.4 In-vehicle networking system

FlexRay is resilient of communication channel errors, sampling signals, consistency checking, encryption, and decoding. The FlexRay bus is a fault-tolerant, modular bus network that provides higher data rates at high speeds than CAN, but due to its higher cost and reliability, it is undesirable to be used in automobiles [16].

2. Controller Area Network (CAN): The CAN bus system is the prevalent standard used by the automobile industry for in-vehicle networking and connectivity for transferring information between the hardware of the different sensors. The reason for its wider adoption is due to its cost efficiency and flexibility compared with other network technologies. However, there are some issues with CAN bus system like Byzantine Generals Problem (BGP) and incapable of handling bubbling failure. There is an alternative of CAN known as CAN with Flexible Data rate (CAN-FD) that uses different data rate during one cycle of information transmission. CAN-FD has a maximum bandwidth of 8 Mb/s [17].

3. Local Interconnect Network (LIN): LIN is also commonly used by the automobile industry due to its lower cost as compared to CAN, due to the use of single wire network and simple software creation. It uses UART ports making variety of microcontrollers capable such as 8-bit microcontrollers used as LIN controllers without special requirements. The LIN network is based on linear bus topology working on master–slave mode and robust against transmission errors. But it has the maximum transmission speed of 20 kb/s, which is very low compared to other bus networks.

2.4 State of the Art of Intelligent and Autonomous Vehicle Technologies

4. Automotive Ethernet (AE): The AE is an innovative technology capable of providing a higher 100 Mbps bandwidth with its supported maximum duplex multi-access connectivity suitable for multimedia and adaptive driving assistance (ADAS). The AE has better security features and low latency compared to CAN and LIN. It employs IP-based routing scheme that prevents the ECU from being hacked and prevents attacker from full control to the whole Ethernet. In case of CAN and LIN, they did not consider the security aspect thoroughly [18]. It has other features such as it requires only less number of ECUs and cables that provide high bandwidth. Due to the use of common Ethernet standards with only slight modifications and its availability helps it for easily adoption in automotive industry.
5. Media Oriented Serial Transport (MOST): The development of the MOST bus is intended to meet the requirement for infotainment systems with higher data speeds in the automotive environment. The MOST is based on daisy chain or ring topology and synchronous data communication to transport multimedia and data signals via plastic optical fiber (POF) (i.e., MOST25, MOST150) or electrical conductor (i.e., MOST 50, MOST150) in physical layers. It is an ideal solution for ADAS system as data is transmitted through optical cable. However, it has some drawbacks such as using MOST in vehicle is expensive as well as a single faulty MOST node causes complete network shutdown.

2.4.2 Connected Vehicle Technology

In Cooperative Intelligent Transportation System (C-ITS), also known as connected vehicle technology largely, depends on VANET for transmitting basic safety message, non-safety messages like multimedia messages along with infotainment messages. Recently, connected vehicles have gained a lot of attention from industry and academia that could potentially help the safe driving, and traffic situation such as congestion, accident, and road construction. The advancement of connectivity in vehicles includes different types of communication technologies for autonomous vehicles. The communication between the vehicles and its surroundings such as neighbor vehicles, pedestrian, RSU is based on inter-vehicle communication. In next subsection, we present the details on the inter-vehicle communication.

2.4.2.1 Inter-Vehicle Communication

The intelligence of the individual vehicle exclusively depending upon the in-vehicle communication based on complex and expensive sensors and control unit inevitably leads to safety concern as well as vehicle control inadequacy. In addition, installing large number of expensive sensors and control equipment in each independent vehicle hinders the adoption of intelligent vehicle due to high cost and limited processing power. To enhance the safety, security, and reliability of the driver and the vehicle and to overcome the processing limitation, inter-vehicle communication is required to assist the in-vehicle communication so that the intelligent vehicle can be introduced in the market.

In inter-vehicle communication, the vehicle can communicate with other vehicles as well as with infrastructure and the cloud using wireless communication. The inter-vehicle communication depends on VANET technology. VANET uses diverse kinds of communication protocols based on the type of communication and type of infrastructure. In VANET, the Wireless Access in Vehicular Environments (WAVE) protocol provides the basic radio standard for dedicated short-range communication (DSRC) operating in the 5.9 GHz frequency band, which is based on the IEEE 802.11p standard [19]. In Europe, an equivalent to DSRC is the ETSI ITS-G5 standard, which is the only commercially available short-range V2X technology known as ITS-G5. The ITS-G5 is based on the IEEE 802.11 standard, i.e., Wi-Fi and the standardization in EU as ETSI EN 302 663 [20]. In addition to this, emerging cellular technologies such as LTE and 5G technology can be used for V2X communication in VANET. Vehicular communications can be achieved in the infrastructure domain for vehicle-to-infrastructure (V2I) communications to connect to roadside units (RSUs) or in the ad hoc domain for vehicle-to-vehicle (V2V) communications or vehicle-to-everything (V2X) communications [21]. All the above-mentioned communication can be carried out through on-board units (OBUs) to connect to RSU in an infrastructure mode or forming mobile ad hoc networks to communicate in a decentralized mode. The first automotive industry to implement the V2X based on DSRC technology in vehicle is Toyota motors in 2016, which was followed by GM motors in 2017. The basic objective of V2X communication is traffic efficiency, road and passenger safety and energy economy. Depending on the type of technology used, there are two types of V2X communication technology as shown in Fig. 2.5. They are 802.11p based and cellular network-based. The 802.11p is based on WLAN technology that works in an unlicensed band, while cellular network is based on licensed band technology that is expensive.

The cooperative driving based on cooperation of DSRC and cellular technology provides additional information sources beyond the information that are sensed by the sensors attached to the individual vehicles. It facilitates cooperative communication so that it can reduce misunderstandings regarding the intended prospective planned movement on the road. It is based on the observed obstacles that are beyond the sensor line of sight and internal vehicle characteristics. It provides higher accuracy and detects fast change in road conditions that minimizes the uncertainties and improves the information quality.

A. **VANET Technology Overview**

i. *Dedicated Short-Range Communications (DSRC)/802.11p Protocol*

In DSRC, the vehicles communicate with each other or the infrastructure using V2V, V2I, or V2X mode using short- to medium-range communication with stability and reliability [22]. It is based on Wi-Fi technology using 8020.11p protocol. It is a part

2.4 State of the Art of Intelligent and Autonomous Vehicle Technologies

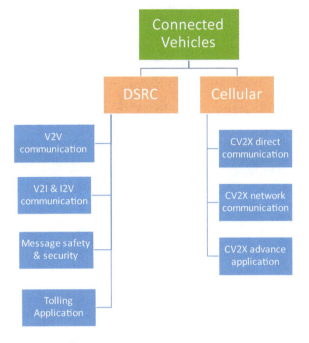

Fig. 2.5 Connected vehicle (V2X) based on DSRC and cellular technology

of radio technology, i.e., IEEE 802.11 WLAN technology known as Wireless Access in Vehicular Environments (WAVE) in USA and ITS-G5 in EU [20]. In the beginning of ITS, DSRC has gain lots of popularity and attention that provides communication between vehicles with high mobility in dynamic network environment. The Federal Communications Commission (FCC) has allocated a 75 MHz spectrum in the 5.9 GHz band for transportation security applications. The 5.9 GHz is split into seven 10 MHz channels. Among them, one channel is reserved for warning and alert messages, which is called the control channel. The others six channels are used for other obligatory services, i.e., service channels, while some channels are reserved as idle for future use. The service channels may also support different data exchange rate as per the requirements.

Owing to the above reasons and features, DSRC technology is used in the following type of applications:

- Safety transportation
- In extreme weather conditions
- For reliable and secure communications
- For low latency and high data rates communication.

Besides safety application, DSRC can also be used for high performance in terms of quality of service (QoS). Figure 2.6 shows the DSRC standard suites [19].

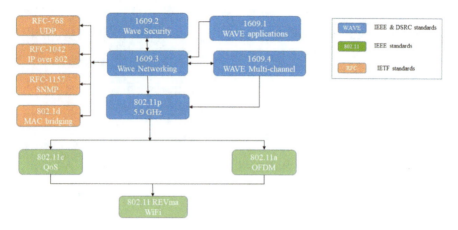

Fig. 2.6 DSRC standard suites

DSRC enabled VANET applications:

DSRC is used in several VANET application due to the advantages mentioned above. The following are some of the application of VANET enabled by DSRC:

(1) *Safety application*: Safety is the utmost priority of the vehicles on the road. The DSRC helps to deliver safety event messages in timely manner. There are 34 various types of safety applications and 11 non-safety applications defined in Vehicle Safety Communications (VSC) projects [23].
(2) *Cooperative Collision Avoidance (CCA)*: Its main goal is to prevent collision between vehicles and avoid crashes during lane change on the road. The CCA safety application will be triggered automatically when there is likelihood of collision.
(3) *Emergency Warning Messages (EWM)*: In case of any emergency event near the accident location, EWM delivers warning messages to the neighbor vehicles to avoid the danger.
(4) *Traffic managements*: In case of traffic congestion, this type of application helps in reducing traffic flow by broadcasting information related to that location in near real time. This application sends the traffic status around the junction or intersection as well as information regarding construction of the pavement to all the vehicles around that location using multihop communication based on DSRC. It provides solutions for solving congestion, reducing fuel consumption, and collision between the vehicles.
(5) *Infotainment applications*: The applications such as advertisements, entertainment, and comfort application mainly provide extravagance and entertainment depending on the user demands. Since these types of infotainment applications consume lots of bandwidth, we need to give priority to the safety application first.

2.4 State of the Art of Intelligent and Autonomous Vehicle Technologies

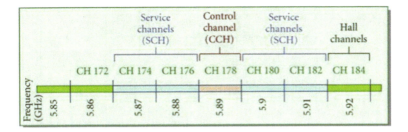

Fig. 2.7 Wave channel arrangement. Adapted from [19]

Some of the examples of these types of applications are electronic toll collection, Internet access for multimedia connection to the cloud, local advertisement based on location, etc.

ii. *Wireless Access in Vehicular Environments (WAVE) Communication*

WAVE is a technology standard established by the IEEE for public safety and ITS service in vehicular communication. WAVE is a short-range wireless communication technology developed based on IEEE 802.11p protocol and IEEE P1609 higher layer safety protocol. It provides message exchange service through short-range wireless communication between a roadside unit (RSU) and an on-board unit (OBU) or between vehicle-mounted terminals. Figure 2.7 shows the WAVE channel arrangement based on control and service channels.

The IEEE 1609 family of standards for WAVE has four standard protocols for vehicular communications. The IEEE1609.1 standard defines the services and WAVE resource manager application in a mobile vehicle environment. The IEEE 1609.2 describes the security services of applications as well as secure message processing and formatting. The IEEE 1609.3 standard describes networking, routing, and transport layer services including the management information base (MIB). The IEEE 1609.4 standard describes the specifications of the multichannel in the DSRC for PHY and MAC layer in IEEE802.11p. The structure of the WAVE layer and its relation to each standard is shown in Fig. 2.8.

iii. *Cellular Technology:*

The intelligent transportation system and IAV gained a lot of attention in the last decade. To facilitate a huge number of application in future vehicular communication such as transportation economy, safety and infotainment of both vehicles and drivers, the IAV requires very low latency, reliability, high level of security, and privacy. The inter-vehicle communication based on VANET for connected vehicles mentioned in the previous section has gained lots of attraction in research and industry. Similar alternatives for connected vehicles have been established and standardized in China, Japan, EU, and other countries. However, those technologies are not enough due to some deficiencies. The connected vehicles based on VANET require substantial infrastructure installations such as RSUs that need lots of investment in the commencement. The IAV communication requirements and applications

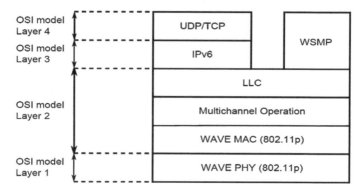

Fig. 2.8 WAVE communication protocol stack

are increasing rapidly while the network configuration, latency, poor scalability, mobility, enormous scale network deployment, security, and achievable data rates of DSRC-based vehicular communication cannot get closer to the ever-growing need of these applications. The DSRC cannot evolve significantly to keep up with the increasing advanced and progressive use cases. This is the main reason for cellular technology to take part in V2X communication research and development. On the other hand, the cellular network is a well-developed matured commercialized technology with scope of future enhancement. There are research and development of using cellular technology supporting vehicular communication known as LTE for vehicle (LTE-V). LTE-V is an alternative technology for ITS accepted by telecom operators and automotive industry, i.e., vendors and manufacturers of automobile technologies. It claims to provide low cost, rapid development, and implementation by using the existing cellular base stations making the urban transportation more efficient and manageable.

Similarly, researchers from automotive industry and academia are working together to enhance the application of autonomous vehicles. This will provide additional potential to the connected vehicles for lower latency, higher reliability, more secure transmission of personal information and high-speed infotainment application as well as real-time event information dissemination.

Recently, emergent connected vehicle technology based on LTE and evolving 5G network to assist the interchange of V2X messages between vehicles and infrastructure is in the rise. It is known as cellular V2X (C-V2X), and it is gaining momentum and support from telecommunication sector and automotive industry. The 5G Automotive Association (5GAA) is a multi-industry consortium that defines the C-V2X technology and its evolution toward 5G. The 5GAA develops test as well as endorse 5G communication solutions that help to accelerate the global market penetration, accessibility, and initiate the C-V2X standardization. The 5GAA objective is to show long-term promise of C-V2X functionality and offers superior performance and ever-increasing capabilities based on 3GPP cellular technology, which is better than IEEE 802.11p. The 5G C-V2X supports a large number of use cases with low latency, high

2.4 State of the Art of Intelligent and Autonomous Vehicle Technologies

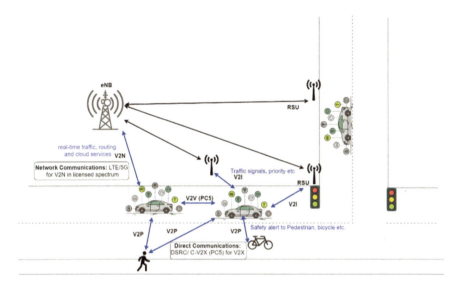

Fig. 2.9 Three different modes of 5G C-V2X based on Rel-14 C-V2X

bandwidth, and reliability. In connected vehicles, the 5G C-V2X provides vehicle-to-network (V2N)-based services including cloud communication. In Europe, the extension with V2N was achieved under the C-ITS platform umbrella [24] with cellular systems and broadcast systems (TMC/DAB +). Figure 2.9 shows the three different modes of vehicular communication, i.e., V2V, V2I, and V2X based on 5G C-V2X on Release 14 C-V2X. The 5G C-V2X can further enhance the safety use cases for future autonomous vehicles by introducing new radio (NR) in 5G C-V2X known as 5G NR C-V2X. The 5G NR is an integrated connectivity framework to expand into innovative connected vehicles in the automotive industry. Recently, Qualcomm designed a reference C-V2X chipset called Qualcomm 9150 C-V2X that is combined with GNSS, which provides a complete solution for trials and commercial development [25]. In Rel-14, the advancement to 5G NR-based C-V2X will extend innovative capabilities for intelligent and autonomous driving supporting backward compatibility. It supports direct C-V2X communication, which has twice the communication range, better demonstration of non-line of sight (NLOS), high reliability, and cost-effective than its counterpart 802.11p. It does not depend on cellular network for safety-related activities for direct communication. The new radio concept in 5G C-V2X provides very low delay, enhance throughput, and ultra-reliable communication abilities. This feature enables progressive and new use cases for IAV.

2.5 Battle for Adoption

The FCC has introduced the DSRC standard more than two decades in the field of intelligent transportation service to provide safety and efficacy of the vehicles on the road. In the USA, the federal government proposed a mandate to implement V2V in new vehicles based on DSRC. Two groups of global automakers and research institutes work toward using wireless communication in autonomous vehicles. One group is attracted toward using DSRC technology in autonomous vehicles while the other groups support the use of C-V2X technology due to its potential and evolution toward 5G. They switched their focus to C-V2X platform and involved in chip making, testing, and demonstration. Since cellular technology is not mature in vehicular communication field compared to DSRC, it might take some time to be widely deployed, as it has not been extensively tested. It is operated in licensed band without dedicated and specific MHz band. The 5GAA and other research institutes like Qualcomm and Huawei back up this technology and advocate for C-V2X adoption because of its advantages in vehicular communication.

V2X standards require a global solution [26]. Recently, there are two standards for connected vehicles with primarily different architectures. It will soon be mandatory to use V2X communications in new vehicles but there is no guideline specifying how the event critical messages should be transmitted. V2X diverged into two different standards, DSRC and C-V2X, with fundamentally different architectures. This makes it difficult for original equipment manufacturers (OEMs) to choose which standard they should select and even more challenge is to synchronize a single global solution. To solve this issue, Autotalks, which is a V2X chipmaker company decided to work on one global V2X solution. They manufactured the world's first dual mode C-V2X chipset (AEC-Q100 grade 2 chipsets) that supports both cellular and DSRC technology [26]. The OEMs can use both C-V2X and DSRC technology in one chip. They mentioned some of the benefits to OEMs as lowest development cost, certification feasibility, single global certification, alignment between V2X and cellular network, and cost-efficient. It is also beneficial to the users as well in terms of safety and security, ensuring the use of safety and infotainment messages and pay as you go scheme while using C-V2X.

2.6 Market Demand of Automotive Cyber Security

There is a serious concern among the vehicle users regarding car hacking after the demonstration was presented the security conferences Black Hat and DEF CON in 2017. The connected cars need to be well-protected against cyber-attacks. The general expectation is that this will lead to an increased demand for solutions that address the most urgent needs, providing improved security for the wireless interfaces and a first level of isolation in the in-vehicle networks. On the longer term, we expect that

2.6 Market Demand of Automotive Cyber Security

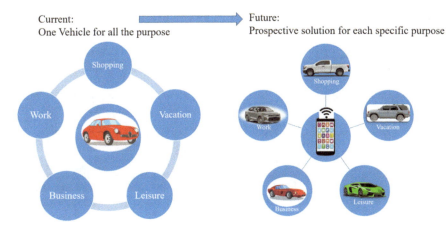

Fig. 2.10 Prospective mobility solution for performing specific purposes

security will become an integral part of the design of the connected vehicle and that the demand for security products will steeply increase therefore.

Traditionally, the automotive industry has been conservative in adopting features offered by consumer electronics. However, the connected vehicle is finally becoming a reality, and it will likely redefine the entire automotive industry. Vehicle manufacturers must find ways to deliver the advanced features their customers demand, into their "smartphones-on-wheels." They will also need to embrace security solutions that are widely used in smartphones and IT infrastructures, but that are relatively new to the automotive world. Examples of such technologies are firewalls, intrusion detection and prevention systems, virtualization technologies, and secure firmware updates.

Consumers currently use their automobiles for all purposes such as going to work, shopping, business travel, leisure activities, and vacation. In future, they choose an appropriate mobility option for each specific purpose for performing various tasks as shown in Fig. 2.10.

2.7 Summary

This chapter focuses on the development of the intelligent and autonomous vehicle, history, and classification of autonomous vehicle driving levels based on SAE and NHTS classification. This chapter provides the latest trends and state-of-the-art intelligent and autonomous vehicles technologies and provides brief information regarding inter-vehicle and intra-vehicle communication. It discusses the technical global status of the autonomous vehicle industry, megatrends, technology adoption battle, market demand, and automotive cyber security.

References

1. K. Sjoberg, P. Andres, T. Buburuzan, A. Brakemeier, Cooperative intelligent transport systems in Europe: current deployment status and outlook. IEEE Veh. Technol. Mag. **12**(2), 89–97 (2017)
2. DoT, *Connected Vehicles*. [Online]. Available: https://www.its.dot.gov/cv_basics/index.htm
3. A. Broggi, A. Zelinsky, M. Parent, C.E. Thorpe, Intelligent vehicles, in *Springer Handbook of Robotics* (2008), pp. 1175–1198
4. R.D. Leighty, *DARPA ALV (Autonomous Land Vehicle) Summary* (1986)
5. E.D. Dickmanns, V. Graefe, Dynamic monocular machine vision. Mach. Vis. Appl. **1**(4), 223–240 (1988)
6. D.A. Pomerleau, *Alvinn: an autonomous land vehicle in a neural network* (Pittsburgh, 1989)
7. V. Barabba, C. Huber, F. Cooke, N. Pudar, J. Smith, M. Paich, A multimethod approach for creating new business models: the general motors OnStar project. Informs J. Appl. Anal. **32**(1), 1–108 (2002)
8. I.C. Society, *Part 11: Wireless LAN Medium Access Control (MAC) and Physical Layer (PHY) specifications: Higher-Speed Physical Layer Extension in the 2.4 GHz Band* (3 Park Avenue, New York, 2000)
9. B. Zhang, ELON MUSK: In 2 years your Tesla will be able to drive from New York to LA and find you, in *Automotive News*, 10-Jan-2016
10. M. Bojarski et.al, End to end learning for self-driving cars. Comput. Vis. Pattern Recognit. (2016)
11. M. Laris, Waymo launches nation's first commercial self-driving taxi service in Arizona, in *Washington Post* (2018)
12. J. Shuttleworth, SAE Standards News: J3016 automated-driving graphic update (2019). [Online]. Available: https://www.sae.org/news/2019/01/sae-updates-j3016-automated-driving-graphic. Accessed 30 Dec 2019
13. P. Godsmark, The definitive guide to the levels of automation for driverless cars. https://driverless.wonderhowto.com (2017)
14. N. Navet, F. Simonot-Lion, In-vehicle communication networks—a historical perspective and review. (Luxembourg, 2013)
15. D. Porter, 100BASE-T1 ethernet: the evolution of automotive networking (Dallas, Texas, 2018)
16. C.P. Quigley, R. McMurran, R.P. Jones, P.T. Faithfull, An investigation into cost modelling for design of distributed automotive electrical architectures, in *2007 3rd Institution of Engineering and Technology Conference on Automotive Electronics* (2007), pp. 1–9
17. BOSCH, CAN with flexible data-rate specification (2012)
18. C. Patsakis, K. Dellios, M. Bouroche, Towards a distributed secure in-vehicle communication architecture for modern vehicles. Comput. Secur. **40**, 60–74 (2014)
19. Y.L. Morgan, Notes on DSRC & WAVE standards suite: Its architecture, design, and characteristics. IEEE Commun. Surv. Tutorials **12**(4), 504–518 (2010)
20. ETSI, Intelligent Transport Systems (ITS); access layer specification for intelligent transport systems operating in the 5 GHz frequency band. (France, 2013)
21. R. Shrestha, S.Y. Nam, Trustworthy Event-Information Dissemination in Vehicular Ad Hoc Networks. Mob. Inf. Syst. **2017**(1), 1–16 (2017)
22. R. Shrestha, R. Bajracharya, and S. Y. Nam, "Challenges of Future VANET and Cloud-based Approaches," *Wirel. Commun. Mob. Comput.*, vol. 2018, no. Article ID 5603518, pp. 1–15, 2018
23. DoT-NHTSA, "Vehicle Safety Communications Project," 2006
24. CAR2CAR, "C-ITS: Cooperative Intelligent Transport Systems and Services," *CAR 2 CAR Communication Consortium*, 27-Aug-2019

25. Qualcomm, "Cellular V2X Overview -Qualcomm 9150 C-V2X Chipset," *Qualcomm*, 2019. [Online]. Available: https://www.qualcomm.com/solutions/automotive/c-v2x
26. Autotalks, "One global V2X solution: DSRC and C-V2X," *Autotalks*, 2019. [Online]. Available: https://www.auto-talks.com/technology/global_v2x_dsrc-and-c-v2x/. Accessed 22 Aug 2019

Chapter 3
Security and Privacy in Intelligent Autonomous Vehicles

3.1 Overview

In this chapter, we mostly focus on cybersecurity and privacy in intelligent and autonomous vehicles (IAV). This chapter starts with the basics of cryptography and then proceeds to different types of advanced security and encryption schemes that can be used in autonomous vehicles. The cyber security in intelligent and autonomous vehicles can be a combination of physical security, information security, security elements, policies, standards, legislation, and risk mitigation strategies. We introduced the updated cybersecurity framework that provides a specific categorization and structural framework for institutions to describe their current cybersecurity position, state for cybersecurity, identify and prioritize security improvements, assess security progress, and plan concerning cybersecurity risks. Then, we discuss the five key technological cybersecurities to protect any company, organization, and IAV against cyber-attacks. A threat modeling method (TMM) is also required to investigate the potential threats so that the IAV system is fully secured from unknown attacks. The TMM is used to defend the cyber-physical system from attackers and detect the threats before they create severe damage. Some of the examples of TMM are STRIDE, PASTA, VAST, etc. The vulnerability is the weak point in the scheme that is misused by the malicious attacker in the form of attacks for their own advantages. We discuss some of the taxonomy that can be found in vehicular system such as autonomous vehicle vulnerability taxonomy, defense taxonomy, and privacy taxonomy.

3.2 Cryptography Introduction

Cryptography is the art and technique for securing the communication of creating ciphertext from plaintext in the presence of adversaries so that the adversaries cannot know the information of the ciphertext. Cryptography along with cryptanalysis are

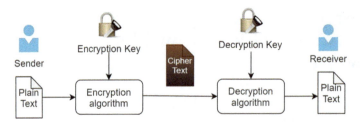

Fig. 3.1 A basic encryption system

the two branches of cryptology. Cryptanalysis is the art of solving and revealing the contents of the code. In the modern context, cryptography and cryptology can be used interchangeably and defined as the techniques for secure communication between two parties. Cryptology is the study of codes for both purposes, i.e., creating and solving them. The purpose of cryptography is to secure digital data by converting the readable information to non-readable information or encrypted format as shown in Fig. 3.1. It provides fundamental information security services based on mathematical algorithms and prevents from the data from malicious activities. A cryptosystem is a set of cryptographic algorithms required to implement any security service. Cryptosystems generally consist of three algorithms; they are key generation, encryption, and decryption. A cryptosystem has the following components:

- Plaintext: The plaintext information should be protected from hackers throughout the information exchange.
- Encryption algorithm: The encryption algorithm converts the input plaintext to a unique ciphertext outputs with the help of encryption key.
- Ciphertext: The encrypted plaintext element is transferred through the server.
- Decryption algorithm: The decryption algorithm reverses the encryption process. For input ciphertext and decryption key, it produces the original plaintext.
- Encryption key: It is the key used during the encryption algorithm to generate the ciphertext from the input plaintext.
- Decryption key: It is the key used to calculate plaintext from the ciphertext.

  A basic encryption system is given below in Fig. 3.1.

3.3 Cryptography Objective

The primary purpose of cryptography is to guarantee the following four basic security services.

3.3 Cryptography Objective

3.3.1 Confidentiality

It is the property, which ensures that the information is not disclosed to any unauthorized entity. It provides identity and location privacy of the user. The data confidentiality can be maintained by using encryption algorithms so that no third party can read the data except the sender and the receiver during the data transmission.

3.3.2 Data Integrity

Data integrity ensures that the data is not modified at the receiver side, therefore ensuring its accuracy and completeness. Data integrity ensures and verifies that the received data is unaltered during transmission from the source to the destination node. Data must be accurate and prevented from modification, corruption, or unauthorized access. The data integrity can be maintained by using data encryption, access control, data backups, data validation that confirms that the data is not corrupted during transmission.

3.3.3 Authentication

Authentication is the process of identifying the authorized entity of the message. It informs the receiver whether the data received was sent to an authorized entity. Authentication services have two types: (a) message authentication: identifies the sender and (b) entity authentication: identifies that the data was received from a specific entity. In addition, authentication provides information on parameters like date and time of creation of the message.

3.3.4 Non-repudiation

It ensures that an entity cannot refuse its ownership to any data or information was created by itself. If there is a dispute over the exchange of data, this property proves to be very useful. For example, if the non-repudiation service is enabled in a transaction, on placing in online order, the buyer cannot deny the purchase order.

3.4 Cryptographic Primitives

Cryptographic primitives are the cryptographic algorithms used to build cryptographic protocols for computer security systems. They are the tools and techniques that are employed to obtain a set of desired security services. They can be classified as shown in Fig. 3.2, and they are mentioned below.

In cryptosystems, the below mentioned primitives are very intricately related and a combination of them is used to obtain required security services. In modern cryptosystems, strings of binary digits need to be processed and converted into other binary strings. Based on the method of processing of these binary strings, symmetric key encryption and asymmetric key encryption schemes can be classified into the following types.

Two main type of encryption algorithms are described below.

3.4.1 Symmetric Key or Secret Key Encryption

Symmetric key encryption uses only one key for both the encryption and decryption purpose. The key plays the main role so for security reasons; secret key should not be revealed to the third party as shown in Fig. 3.3.

Substitution and permutation are two main techniques used in most secret key algorithms for encryption. Substitution is the term used for the mapping of a set of values to another. While permutation is the term used for reordering the bit positions of every input. The iterative use of the aforementioned two processes is called rounds. As the number of rounds increases, the algorithm becomes more secure.

Since both the parties should have the secret key with them, this turns out to be a major problem with secret key cryptography. This is because transmitting the key over a secure channel is not easily achievable.

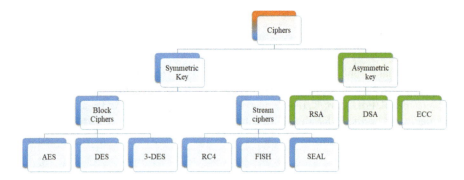

Fig. 3.2 Encryption system classification

3.4 Cryptographic Primitives

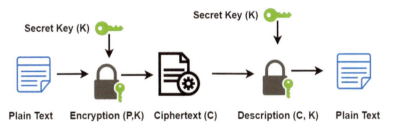

Fig. 3.3 Secret key encryption

Types of symmetric encryption schemes are:

1. *Block ciphers*: In block ciphers, the plain binary text is processed in either blocks or groups of bits at a time. The size of the block is fixed for a specific algorithm. For example, the DES scheme has a block size of 64 bits while the AES scheme has a block size of 128 bits. Several block cipher schemes are used for encryption. Most popular and prominent block ciphers [1] are described in Table 3.1.
2. *Stream cipher*: In stream ciphers, plain binary text is processed bit by bit, i.e., a series of operation is performed on a single bit of plaintext and one bit of ciphertext is generated. We can see stream cipher as a block cipher with blocks of size one bit each. The most common stream ciphers [2] are given in Table 3.2.

Table 3.1 Type of block cipher schemes

Features	Advanced encryption standard (AES)	Data encryption standard (DES)	Triple data encryption standard (3-DES)
Introduced	2000	1977	1978
Ciphertext	Symmetric block cipher	Symmetric block cipher	Symmetric block cipher
Block size	128, 192, 256 bits	64 bits	64 bits
Key length	128, 192, 256 bits	56 bits	168 bits (3-key) 112 bits (2-key)
Security	Considered secure	Not secure enough	Inadequate
Cryptanalysis resistance	Strong against differential and linear cryptanalysis	Vulnerable to differential and linear cryptanalysis	Vulnerable to differential, brute-force attacker can analyze plaintext
Possible keys	$2^{128}, 2^{192}, 2^{256}$	2^{56}	2^{112}
Encryption and decryption	Faster	Moderate	Faster than DES

Table 3.2 Types of stream ciphers schemes

Features	Rivest Cipher 4 (RC4)	FIbonacci SHrinking (FISH)	Software-Optimized Encryption Algorithm (SEAL)
Introduced in	1987	1993	1997
Key length	8–2048	Variable	–
Initialization vector	No initialization vector in RC4	–	32 bits
Internal state	2064 bits	–	–
Attack	Shamir initial-bytes key-derivation	Known plaintext attack	–
Computational complexity	213 or 233	211	–

3.4.2 Asymmetric Key or Public Key Encryption

In this scheme, a pair of keys is used for encryption and decryption algorithms. The public key is derived from the private key and the two keys should be different, however, they are mathematically related. One of the keys is called public key, as everyone knows it, while the other one is called private key only known to the owner. Therefore, asymmetric key cryptography is also known as the public-key cryptography as shown in Fig. 3.4.

Properties of asymmetric key encryption algorithm are listed below.

- Every user must have a pair of different keys, private key, and public key. One meant for encryption of the data while the other meant for its decryption.

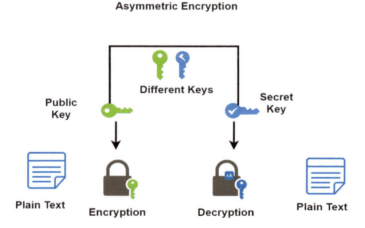

Fig. 3.4 Asymmetric key encryption scheme

3.4 Cryptographic Primitives

Table 3.3 Comparison of the common asymmetric encryption schemes

Features	Rivest–Shamir–Adleman (RSA)	Digital signature algorithm (DSA)	Elliptic curve cryptography (ECC)
Introduced in	1977	1991	1985
Introduced by	Ron Rivest, Adi Shamir, and Leonard Adleman	National Institute of Standards and Technology (NIST)	Suggested by Neal Koblitz and Victor S. Miller; endorsed by NSA
Key length	2048 bits	2048 bits	256 bits
Security	Considered secure	Not considered secure	stronger security and increased performance
Based on	Integer factoring	Discrete logarithm	Elliptic curve discrete logarithm
Possible attack	Brute-force attacks, side-channel analysis	Brute force, side-channel analysis, timing attacks	Quantum computing attacks, invalid curve attack

- The public key is put into public repository while the private key is stored as a secret.
- Though public and private key are mathematically related, it is computationally not possible to get one from another.
- When a sender needs to send data to the receiver, the sender obtains the public key of the receiver from the repository, encrypts the data, and then transmits it. The receiver uses his private key for decryption.
- The process of encryption and decryption is slower due to the key length. Hence, this scheme is slower than the symmetric key scheme.

The comparison of the common asymmetric encryption schemes [3] is given below in Table 3.3.

3.4.3 Digital Signatures

A digital signature is a technique that associates an entity with the digital data. The digital signature scheme is based on public-key cryptography. This association is independently verifiable by the receiver as well as any third party. Digital signature is a cryptographic value, which can be calculated from the data and some secret known to the signer.

1. Encryption with Digital Signature

As the public key of the user is available in the open domain, so, anyone can spoof his identity and send any message to the receiver. This makes it important for users employing public-key encryption to use digital signatures along with encrypted data to be assured of message authentication and non-repudiation. This can be achieved

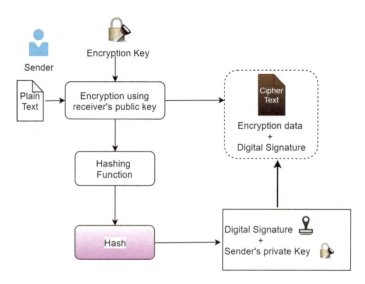

Fig. 3.5 Encryption technique

by combining digital signatures with an encryption scheme. There can be two cases, either to digitally sign first or to do encryption first. In cryptosystem based on sign first and then encrypt, it has one limitation that is the receiver can exploit this scheme by spoofing the identity of the sender and use it to cause manipulation. Hence, this scheme is not preferred. The process of encrypting and then sign is more reliable and widely adopted as illustrated in Fig. 3.5. The receiver first verifies the signature by using sender's public key after receiving the encrypted data and signature on it. After the validation of the signature is ensured, the receiver then uses his private key to decrypt the received data.

3.4.4 Homomorphic Encryption

The traditional secret key sharing encryption schemes are not completely secure for critical information. While using the third-party service provider, the plaintext may be leaked throughout the data processing during the encryption and decryption algorithm. Thus, homomorphic encryption (HE) is required to maintain the privacy of the user's text. HE is an encryption scheme that allows users to encrypt their information using cryptographic keys. It permits the third party to accomplish specific type of mathematical operations on the encrypted information without decryption, while maintaining the privacy of the users' encrypted information [4]. In HE, the mathematical operation on the plaintext during encryption is equivalent to another

3.4 Cryptographic Primitives

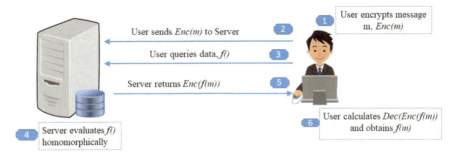

Fig. 3.6 Homomorphic encryption scenario

operation performed on the ciphertext. The common homomorphic operation on the plaintext with a corresponding ciphertext operation are as follows:

Let $E(x_1) = x_1^e$ and $E(x_2) = x_2^e$, then

Addition homomorphism:

$$E(x_1) + E(x_2) = x_1^e + x_2^e = (x_1 + x_2)^e = E(x_1 + x_2)$$

Multiplication homomorphism:

$$E(x_1) \times E(x_2) = x_1^e \times x_2^e = (x_1 \times x_2)^e = E(x_1 \times x_2)$$

In HE, there are four procedures including the evaluation algorithm [5] as shown in Fig. 3.6. Let the input message tuple $(x_1, x_2, \ldots x_n)$ with plaintext as $P = \{0,1\}$ and Boolean circuit as C [4]. The HE process is as follows:

- $Gen(1^\lambda, \alpha)$ is the key generation algorithm with three output keys along with evaluation key (sk, pk, evk),

$$(sk, pk, evk) \leftarrow KeyGen(\$)$$

- $Enc(pk,x)$ is the encryption algorithm that encrypts a message (x) with the public key (pk) and outputs a ciphertext c,

$$c \leftarrow Enc_{pk}(x)$$

- $Eval(evk, C, c_1, c_1, \ldots..c_n)$ is the evaluation algorithm that generates evaluation output by using evk key as i/p, a circuit $C \in C$ and tuple of i/p ciphertext, i.e., $c_1 \ldots c_n$ and previous evaluation results,

$$c^* \leftarrow Eval_{evk}(evk, C, c_1, c_2, \ldots c_n)$$

- *Dec(sk, c)* is the decryption algorithm that decrypts a ciphertext or evaluation output (*c**) using the secret key (*sk*). It generates the actual message *x* as the output,

$$x \leftarrow Dec_{sk}(c).$$

3.5 Cyber Security in Intelligent and Autonomous Vehicles

There is a security threat and cyber-attack in hardware such as computer networks and information system such as data, programs, and other services due to unauthorized access and weak protection against malicious entities. So, the cyber security in intelligent and autonomous vehicles can be a combination of physical security, threat modeling, information security, policies, standards, legislation, and risk mitigation strategies.

Figure 3.7 shows the security elements of the intelligent and autonomous vehicles in terms of physical and cyber security. The figure shows the vulnerable parts of the intelligent vehicles that are easily targeted by the malicious attackers by using different means at different levels and their impacts on the vehicle's physical and cyber security [6]. The vehicles security elements are easily targeted by the attacker if proper prevention cannot be taken efficiently. The physical attack only affects the vehicle and its properties; however, the cyber-attack on the intelligent vehicle might have serious consequences that might not only lead to the privacy and identity theft but also may cost the life of the driver, if the attacker manipulates the traffic conditions by changing the directions. The vehicle elements can be secured by isolating different components by their functions, securing their interfaces with other components,

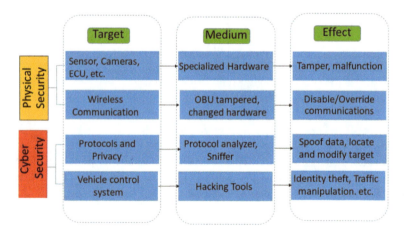

Fig. 3.7 Security elements of autonomous vehicles

3.5 Cyber Security in Intelligent and Autonomous Vehicles

safe guarding the in-vehicle networks and communication channels by adopting cryptographic techniques such as IDS and firmware updates.

We need to consider that the malicious nodes are capable of breaking the IAV security. The malicious nodes can tamper the onboard unit (OBU) and other in-vehicle sensors and can intercept message exchange between the vehicles in in-vehicle communication. The malicious nodes can be an insider attacker that may attack both in-vehicle and inter-vehicle communication. There are several attacker models that have been demonstrated.

3.5.1 Cyber Security Framework

In 2019, National Standards and Technology Institute (NIST) has launched the updated version 1.1 of cyber security framework (CSF) for improving critical cyber security infrastructure and to help institutions on their path toward the development of secure computer systems as shown in Fig. 3.8 [7]. The framework is split into five parts, covering identifying risks and vulnerabilities, protecting and maintaining critical infrastructure, detecting cyber risks as early as possible, responding adequately to attacks and recovering as rapidly and effectively as possible. The updated framework provides a specific categorization and structural framework for institutions to describe their current cyber security position, state of cybersecurity, identify and prioritize security improvements, assess security progress, and plan regarding cybersecurity risks. The framework is not a common suitable solution to every field for critical infrastructure management of the cybersecurity challenges. Every field will have specific challenges and threats, different flaws, and varying tolerances of hazards. Likewise, in case of vehicular networks, the frameworks should be adopted

Fig. 3.8 NIST cybersecurity framework 1.1 including five functions

according to the vehicular network challenges, threats, attacks, and environment. The five functions of the NIST framework ver. 1.1 are briefly discussed as follows:

1. *Identify*: The data, equipment, processes, and resources that allow the institution or system to accomplish functions are identified and handled in accordance with their relative value to the institutional priorities and its risk strategy. Institutional threats, including internal and external ones, are identified and documented. The identified hazards, vulnerabilities, likelihoods, and effects are used to determine the risk. Risk assessment and management are carried out, and risk responses are identified and prioritized.
2. *Protect*: It is necessary to develop and implement adequate protection to ensure delivery of critical infrastructure resources to protect the system from cyberattacks. The identities and credentials for approved devices, users, and processes are issued, managed, handled, checked, revoked, and audited. The user equipment and other assets are authenticated (e.g., single-factor, multifactor) according to the transaction threats (e.g., security and privacy risks for entities, and other institutional risks). Policies, protocols, and practice for the safety of information systems are preserved and used for administration. In accordance with relevant protocols, processes, and arrangements, the strategic security solutions are handled to ensure the security and stability of the networks and its facilities.
3. *Detect*: In the detection step, anomalous behaviors are easily identified and the potential consequences of incidents are known before the occurrence of a cyberattack event. To detect cybersecurity incidents and check the efficiency of protective measures, the information system and properties are monitored continuously. Detection process and methods are maintained to ensure prompt and effective detection of anomalous incidents.
4. *Respond*: In case of respond step, appropriate actions are taken by implementing secure activities based on the detected cybersecurity event. Response procedures and mechanisms are taken to ensure timely response to observed cybersecurity incidents. Analysis is carried out to guarantee appropriate response and to facilitate restoration initiatives. The new response activities are improved by combining knowledge gained from current detection/response activities and previous ones. The response activities include analysis and mitigate the cyber threat as they are detected.
5. *Recover*: In recovery step, the recovery measures and techniques are carried out and managed to guarantee that the networks or properties damaged by cybersecurity incidents are recovered in less time. The recovery plan is enforced before or after an incident with cyber security. Similar to the respond step, the recovery activities such as recovery process and planning are improved by combining knowledge gained from current activities into future activities. The recovery operations should be communicated with domestic and external participants like ISPs, attack device owners, perpetrators, other ISPs and service providers.

3.5.2 Cybersecurity Layers by Design

The security of a vehicle should be considered, including its feature specification from its very inception and during the design stages. The security by design should be in the first place to evade the security risks. Addressing security problems earlier guarantees the vehicle security systems and equipment as well as reduces risks of vulnerabilities at later stages. Cybersecurity must be implemented into the design instead of including at the end of the development phase. Therefore, incorporating cybersecurity into the design requires a suitable cybersecurity layer. The cybersecurity layers achieve defense in depth following the principle of least privilege cyber security. Some of the popular threats to cybersecurity includes viruses, worms, spyware, and ransomware. However, cyberattack strategies are constantly evolving and it is difficult to prevent the security system from these cyber-attacks by implementing only one layer. This implies a need for continuous monitoring and multiple security layers. A multilayered solution is designed to reduce attackers' resources and entry points. The concept employs multiple layers instead of having only one layer of defense, making it harder for cyber attackers to execute successful attacks. There are five key technological cybersecurity layers that need to be considered to protect any company or organization against a cyber-attack as shown in Fig. 3.9.

The summarized five layers of security layers providing robust security protection are as follows:

1. Perimeter: The perimeter layer secures interfaces, such as a tamper-proof trust anchor, which link the vehicle to the outside world. The objective of this layer is to secure the external connection of the IAV by limiting physical access to the vehicle control system to prevent the hackers from hacking the system. The first protective layer provides protection to the telematics control unit (TCU) or onboard diagnostic (OBD) port by adding a secure feature to ensure maximum security.
2. Network: The network layer controls access from external interfaces like Internet to the vehicle's internal network by using a firewall. It separates the networks physically and electronically using the central gateway with firewall. If the hackers get access to the network, then they can spoof the messages and take

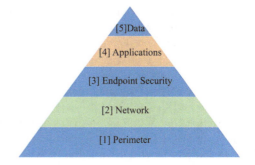

Fig. 3.9 Cybersecurity layers by design

control of the vehicle system. The gateway firewall isolates the network and allows only legitimate devices to communicate with other nodes. One of the important functions of firewall from the security point of view, besides isolation is that firewall separates the external network interfaces from the safety-critical internal vehicle network.

3. Endpoint security: The endpoint security delivers secure connection between electronic components. This layer prevents communicating nodes against various in-vehicle networks data theft. This layer includes the intrusion detection and intrusion prevention schemes. This layer secures the networks with bus control and secure transceiver with cryptographic capabilities for message authentication. It is responsible for access management and keeps communication private and reliable by securing the transceivers.

4. Application: This layer ensures the software running on the processor is authorized, legitimate, and trustworthy. It secures the microprocessor unit (MPU) and microcontroller unit (MCU) as well as responsible for the secure updates of the security features of the vehicles.

5. Data: The data layer is the fifth layer of the multilayered security system of the vehicle system. The data security is very important because it contains all the information about the user, including his identity, bank accounts, medical history, his whereabouts, etc. The integrity, confidentiality, and reliability of the data of the user must be guaranteed. This layer provides the security of the user or vehicle information.

The detailed cybersecurity layers for vehicular system are presented in Chap. 4.

3.5.3 Threat Modeling Method (TMM)

There are several kinds of hacker and attackers in IAV. In fact, there are various attackers, with different motivations, skill levels, and resources. For example, there may be academic researchers who try to take partial control over the vehicle, for scientific reasons. In addition, there may be organized criminals with large budgets, who want to steal valuable data from a vehicle, for financial gain. However, the threats do not only originate from third parties, but also from the user himself by tuning the internal car circuits to unlock extra features or improve engine performance. Furthermore, different attack vectors may be mounted directly from the in-vehicle electronic system. Therefore, a threat modeling method (TMM) is required to investigate the potential threats or attacks so that the IAV system is fully secured from unknown attacks. The TMM provides a structured method to protect the vehicle software from potential attackers, including their intention and goals that is based on attack types and libraries [8]. The TMM is used to defend the cyber-physical system (CPS) from attackers taking advantage of the system's weakness and detect the threats before they create severe damage. There are several types of TMM, but not all of them work for a specific threat. The decision on which type of TMM to use should be based

3.5 Cyber Security in Intelligent and Autonomous Vehicles

Table 3.4 Features and objectives of different types of TMM

TMM	Usage	DFD	Feature	Objective
STRIDE	Medium	Yes	Mature technique	Software-centric
PASTA	Hard	Yes	Risk management	Risk-centric framework attacker-centric perspective, asset-centric output
LINDDUN	Hard	Yes	Data security	Privacy-centric
Attack tree	Easy	–	Security decision	Attack-centric
VAST	Medium	Yes	Risk management	Application-centric
PnG	Easy	–	Low false positives and high consistency	Human attacker-centric

on the requirement of the project. The TMM works well if it is used early during the system design phase. Two or more TMM can be combined together to create a strong view o f potential threat.

In this section, we will briefly discuss only specific type of TMM that is useful in autonomous vehicles, according to the evaluation done by the team of CMU researchers [9]. The list of evaluation criteria for selecting the candidate TMM is strengths, weakness, applicability, adoptability, automation, and tailor ability. Table 3.4 lists the features and objectives of different types of TMM. The potential TMM based on the above-mentioned criteria is as follows:

1. STRIDE: The STRIDE stands for spoofing of personal identity, tampering information, repudiation, information revealing, denial of service, and elevation of privilege. STRIDE helps to identify and classify the threats to the vehicular system. It was adopted by Microsoft in 2002 and has evolved over time; however, Microsoft developed and use another threat model named damage potential, reproducibility, exploitability, affected users, and discoverability (DREAD) [10]. It is the most popular and commonly used TMM in CPS and is a software-centric model. STRIDE can be applied all entry points to identify the threats to the vehicle including threats from hardware attacks, e.g., tampering of onboard units, local storage, etc. After identifying the threats and vulnerabilities, STRIDE threat model can be used as a reference to know how the threats directly affects each of the vehicle assets identified earlier. The STRIDE threat model is shown in Fig. 3.10. The STRIDE threat includes:

    Spoofing of personal identity—Illegal use of the victim's authenticated information to access the device.
    Tampering information—maliciously modifying the information of the device
    Repudiation—Deny performing a malicious action. Non-repudiation is the ability to counter the repudiation attacks.
    Information revealing—Disclosing the user confidential information to public
    Denial of Service—Temporarily disable the device and/or deny services to the user, which can threaten the availability

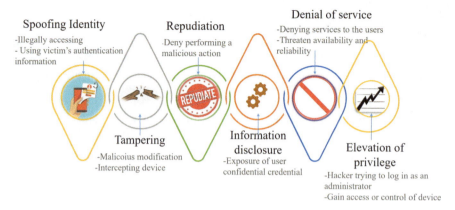

Fig. 3.10 STRIDE threat model

Elevation of privilege—Hacker login to the system as an administrator

2. PASTA: PASTA stands for process for attack simulation and threat analysis. It is a seven-stage threat model framework developed by Tony UcedaVelez in 2012. It is a risk-centric framework with attacker-centric perspective, which produces asset-centric output. Most of the TMM that we are going to discuss are based on data flow diagrams (DFDs). DFD is a part of the design phase of the development cycle that helps to identify system objects, boundaries, and events in the system. PASTA uses DFD in application decomposition layer. The PASTA utilizes risk and impact analysis that overcomes the weakness of LINDDUN and STRIDE.
3. LINDDUN: LINDDUN stands for linkability, identifiability, non-repudiation, detectability, disclosure of data, unawareness, and non-compliance. It has six methodical steps that provide data security and privacy of information. It uses DFD for data flow of the system, and it iterates over all model elements to analyze and detect different types of threats. However, in LINDDUN, the system complexity increases as the number of threat grows rapidly that impacts the adversely on the general threat detection.
4. Attack trees: In 1999, Bruce Schneider introduced the attack trees that depicts the diagram of attacks in tree form. The root of the tree is the goal while the branches are the methods to achieve that goal. The attack trees are used for security decision based on the evaluation of the specific type of attacks on the branches. It is easy to understand, use, and adopt as well as it can be combined with other methods such as STRIDE and PASTA.
5. VAST: The VAST stands for visual, agile, and simple threat that is based on an automated threat-modeling platform called threat modeler. The VAST is scalable and applicable in big organizations that is based on application and operation threat models. The operational threat models use DFDs with an attacker's perspective. It contributes to the risk management directly.
6. Persona non Grata: The Persona non Grata (PnG) is a threat model that is based on the activity of the real human attacker. The PnG helps the security person

3.5 Cyber Security in Intelligent and Autonomous Vehicles

to analyze the system's vulnerabilities and weak points that assist in the early detection of the attacks. It helps in risk management as it is easy to adopt, but it can only detect a limited number of threats.

Adi et al. adapted the two threat modeling process TARA and STRIDE and implemented in the automotive industry for connected vehicles [11]. They presented the list of the most exposed areas of autonomous vehicles, with low-, medium-, and high-risk exposure. The verification is done using AUTOSAR and dedicated hardware. Similarly, Wenjun et al. implemented a connected vehicle threat model based on a tool called SecuriCAD and performed connected vehicle attack simulation [12]. The SecuriCAD is a threat modeling and risk management tool that helps user to model networks. The simulation engine is used to predict different attack success probability. The SecuriCAD attack simulation shows that the DoS attack on a drivetrain network is at high risks as shown in Fig. 3.11. The simulation results show that the firewall is the most vulnerable object found in the vehicle during the DoS attacks. It also shows that the attack can be prevented by improving the security rules of the *FirewallKnownRuleSet*.

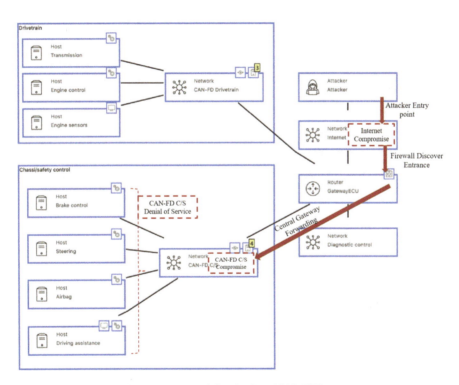

Fig. 3.11 Generalized vehicular threat modeling by SecuriCAD [12]

3.5.4 HARA and TARA Safety and Security Methods

The advancement of intelligent and autonomous vehicles poses cyber security challenges and hazards to road and traffic safety. Securing automotive networks is one of the biggest challenges confronting the automobile industry experiencing a dramatic transition. The security engineering is required for the safety and security of the autonomous vehicles, and this can be achieved by threat modeling technique for security and risk analysis. The automotive functional safety standard organization ISO 26262 defines the road vehicles functional safety engineering and safety standards. It also defines hazards due to the automotive malfunction and introduced a method called hazard analysis and risk assessment (HARA). The HARA assessment is done by considering consequences of failure. It identifies hazards based on information on potential causes. The safety requirements can be divided into functional safety requirements and safety integrity requirements. Functional safety can be accomplished when each specified safety task is performed and the performance level needed for each safety function is met [13]. This is usually done by a method that involves following steps:

- Safety Function Identification: It is necessary to identify the required safety functions and threats.
- Safety function assessment: The Safety functions are required to access and reduce the risks; this involves the safety integrity level (SIL). A SIL refers not just to a feature or part of the system but an end-to-end safety function of the system. The SIL is designed to reduce the relative risk or specify the target level of risk reduction specified by a safety function. The European functional safety standard is established on the IEC 61508 specification that describes four SILs where the most reliable is SIL 4 and the least reliable is SIL 1. The automotive functional security standard (ISO 26262) requires at least automotive security integrity level (ASIL) specific safety features. ASIL is a risk classification method that is an adaptation of SIL. ASIL is used for safety measures specified to reduce the risk associated with potential hazards to an acceptable level.
- Safety function verification: Safety function verification validates if the assigned SIL meets the safety function. The verification function is carried out to decrease or eliminate the deviation to obtain acceptable safety level. The software accuracy, latency, and synchronization of vehicle dynamics should be validated to reduce safety hazards.
- Failure mode and effects analysis (FMEA): The critical or hazardous condition is defined from the FMEA and the failure mode effects and critical analysis (FMECA) of the system under test. Hence, FMEA is best tailored to new developments and improvements in goods and processes. The risk management allows for identifying key components and setting goals in error prevention. The safety engineering standards integrate FMEA and fault tree analysis (FTA) in the automotive domain.

3.5 Cyber Security in Intelligent and Autonomous Vehicles

- Functional safety audits activity: Investigate and analyze the proof that the correct life cycle management methods were systematically and extensively applied in the related life cycle phases.

The HARA process can be used for threat analysis that can be extended and adapted in cybersecurity field. In parallel view to HARA, the principle of safety requirements can be extended to security requirements, and therefore, an equivalent cybersecurity process is established for threat analysis and risk assessment (TARA) to identify possible cybersecurity attacks and risks linked to these attacks with security goals and functional requirements. TARA is based on ISO/IEC 27001 standards, which is an information security standard that manages the asset security. The TARA distinguishes and selects potential threats against the target vehicle that leads to security instances. Its objective is to protect the assets and prevent the system from threats and attacks. In concept phase, the emphasis is on the relationship between HARA and TARA and their coordination in defense system. In ISO 26262, automotive safety integrity level (ASIL) is used for safety measures specified to reduce the risk associated with potential hazards to an acceptable level [14].

Figure 3.12 shows the vehicles risk identification and classification based on functional safety and cyber security in terms of HARA and TARA methodologies.

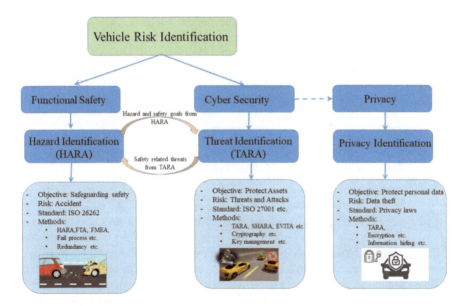

Fig. 3.12 Classification of HARA and TARA safety—cybersecurity methods

3.5.5 Security and Privacy Threats in Vehicular Networks

There are several possible security attacks in intelligent autonomous vehicles, which are listed below:

- **Fake information attack**: The selfish attacker transmits fake information for his own benefits. For example, the attacker may transmit an emergency vehicle approach warning for his way on the road to be clear.
- **Illegal preemption attack**: During an emergency, if roadside unit (RSU) controls the traffic light then the attacker may lawlessly break out the traffic light via RSU for his own needs like the fake information attack.
- **Message replay attack**: The legitimate messages that were already sent by a valid source are sent again by fraudulent or malicious intention or with some delays to interfere the traffic.
- **Integrity**: It is one of the useful properties in securing the IAV. Integrity assures that the messages exchanged between two vehicles must not be tampered or go through any modification.
- **Non-repudiation**: Non-repudiation ensures that any message sent by a legal vehicle cannot deny the message or contents send by it.
- **Access Control**: Access control ensures reliability and security in the system by assigning responsibilities roles to the nodes. The roles are assigned to the authentic users of the system. For safety of the legal entities in the network, the misconducting entities are invalidated from the network.
- **Privacy**: Privacy ensures protection of personal information from illegal users. In IAVs, conditional privacy preservation is revealing the real identity of a vehicle to only authorized users of the network and concealing the real identity to other vehicles and infrastructure like RSU.

3.5.6 Autonomous Vehicle Cyber Security

Security should be the first priority in developing intelligent self-driving vehicles providing the highest standards, quality, and reliability. There is always a threat to the security of the autonomous vehicle since its creation. In other words, manufacturers can no longer solely rely on the physical protection offered by the vehicle's chassis. The simple and most common security issue is vehicle theft and anonymity problem based on vehicle identity number (VIN) as well as fake use of license plate. In order to tackle this security threat, an effective security system was used which is based on the combination of central locking system, alarm, electronic/mechanical immobilizer, and tracker. However, with the advancement of technology, these security systems are not enough. The range of attacks that an IAV faces is extensive and diverse: it varies from relatively simple attacks, in which, for example, malicious messages are sent to a vehicle, to more sophisticated attacks in which hackers may open up ECUs and try to reverse engineer their microcontrollers and software.

Finally, the impact of a successful hacking attack may also widely differ. In certain cases, a hacker may target a specific vehicle, causing limited damage to that vehicle only. However, a hacker may also find an exploit that can be abused over complete control of vehicle. When attacks are easily reproduced by others, the hacker publishes tools and instructions on the Internet. Due to this, the attack impact and the financial damage are much larger. For example, a large-scale attack at random vehicles could easily have an economic impact, because it has the potential to interrupt the traffic in a large geographical region. In addition, the costs for vehicle manufacturers could be high, because of potential recalls and associated brand damage.

1. Autonomous Vehicle Vulnerability Taxonomy

The big challenge for vehicle manufacturers is to implement solutions to prevent vehicle from wide variety of attacks by the hackers, with different motivations, resources, skill levels, and using many different attack vectors, in a cost-effective way. Security threats form a predominant undiscovered area for the automotive industry. Outside the automotive industry, standardized frameworks are used to get the customers in confidence by informing them that their security demands are protected, and they can easily trust the attributes of the product. However, the security researchers are successful in hacking the control system of Jeep Cherokee using the wireless connectivity of the vehicle's entertainment system [15]. They compromised the CAN bus as well as other control functions such as braking systems and steering, by using the wireless communication like Bluetooth and Wi-Fi as attack vector. This shows that the security systems in the vehicles have not be properly implemented that might cause vulnerabilities in the vehicle control system. Therefore, it is essential to secure and protect the vehicle CAN bus system from vulnerabilities that send critical system information to other modules. The vulnerability is the weak point in the scheme that is misused by the malicious attacker in the form of attacks for their own advantages. The autonomous vehicle vulnerability taxonomy is shown in Fig. 3.13.

The attack vectors are based on physical access and remote access. The attack vectors facilitate hackers to exploit the vehicular system vulnerabilities and gain access to the network to perform malicious activities. The attack vectors include Web sites, viruses, mail attachments, pop-up windows, SNS messages, chat rooms, and fake information. It is categorized into invasive and non-invasive methods. In case of invasive method, the attacker uses the onboard diagnostics (OBD2) port to interface with ECU and inject malicious code that gives the attacker access to unauthorized functions [16]. Similarly, the invasive attack may cause serious issues such as packet sniffing, fuzzing and spoofing that leads to falsifying information used by the vehicle's system. An unauthorized user can attack the system and remotely access the vehicle based on user's devices such as smartphones that are coupled to the infotainment system via Bluetooth or Wi-Fi technology, from external devices in proximity such as other V2X-equipped vehicles, or from the cloud.

It can target the critical objects and spoof the GPS information and provide false location information as well as cause jamming to the sensors that disrupt the vehicle sensors from getting critical information. The attacker can target different sensors

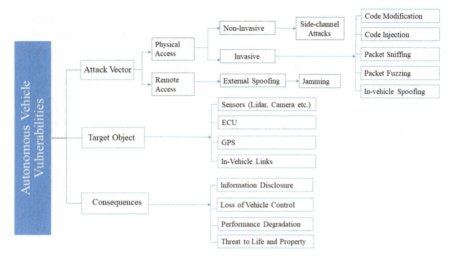

Fig. 3.13 Autonomous vehicle vulnerability taxonomy

such as LiDAR, radar, camera, ECU, GPS, and other in-vehicle communication system that can disrupt the proper functioning of the autonomous vehicle.

The consequences of the attacks are private information disclosure of the user. The attacker might take over the full control of the vehicle and the user cannot control the vehicle, which might have serious life threatening situation as well as loss of property.

2. Autonomous vehicle defense taxonomy

Various automotive manufacturers, academia, and security companies propose the different solutions to combat numerous attacks and secure the in-vehicle networks, which is given in Fig. 3.14. The preventive countermeasures such as device and user authentication thwart the unauthorized users to access the in-vehicle system. Software security based on cryptographic algorithm such as encryption, signature, and digital certificate can prevent the ECU and CAN network from the attacker [17, 18]. The automotive manufacture can use asymmetric key cryptography for secure communication between different ECUs in the in-vehicle system. The firewall can be used to prevent the in-vehicle network by filtering harmful messages and blocking the malicious traffic sources in untrusted environment. In active defense system, continuous monitoring system is installed that keep track of all the logs and records all the activities of the system. Due to the vulnerabilities in the in-vehicle communication system, the attackers easily launch critical attacks remotely. So, continuous monitoring systems such as adaptive security and honeypots can be used to capture the log of the attacker and trace their activities that are vulnerable to the system. It can detect such malicious activities and can take proper action to prevent the in-vehicle system [19]. In case of passive defense system, Intrusion Detection System (IDS) and antimalware can be used to detect and eradicate the malicious

3.5 Cyber Security in Intelligent and Autonomous Vehicles

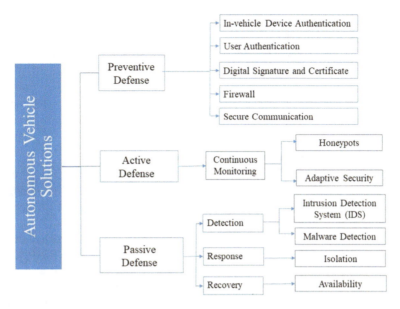

Fig. 3.14 Taxonomy of defense system in VANET

code that has been injected by the malicious attacker into the ECU. In addition, each component should be isolated and independent from each other so that the attacker cannot easily hack into the system that are interconnected with each other. The isolation of the component ensures defense against harmful attacks from the attacker that might compromise the whole system. Moreover, the defense system should be recoverable from the attacks and should provide the trustworthy information at any time to the system for tracing back.

3.5.7 Connected Vehicle Security

The wireless interfaces in connected vehicle imposes the biggest security risks, because they open the door for remote attacks. The attacker does not need to be in direct proximity of the vehicle to gain access to its internal systems, the attacker can access the system remotely by breaching the wireless interfaces.

3.5.7.1 Components of IAV

The components of IAV can be classified into three categories based on the infrastructure, message content, and vehicular system. The classification helps in security assessment and risk analysis. The three categories are as follows:

i. Infrastructure: It consists of central entities, RSU, and other third parties such as transportation authority.
ii. Information content: It consists of different types of messages such as beacon, safety, and infotainment messages that help safe and secure driving of the IAV on the road.
iii. Vehicular system: It consists of vehicular nodes, driver and communication networks.

3.5.7.2 Vehicular Ad Hoc Networks Vulnerabilities

The IAV are vulnerable to different types of attacks where the malicious attacker exploits the vehicular communication network for their own benefit. The vulnerabilities are classified as shown in Fig. 3.15. The Vehicular Ad hoc Networks (VANET) is vulnerable to physical access by the attackers that has direct risk on the vehicle's wireless communication capabilities as well as its sensing systems. The user's privacy might be at risk if secure credentials are not used while exchanging information using wireless communications. Similarly, attackers might exploit the system to launch various attacks based on insecure cryptographic algorithm and inefficient key management. The sensitive messages like user information and safety messages are very important and should be secured all the time by using encryption mechanism. If information is not encrypted properly and there is weak integrity checks then the

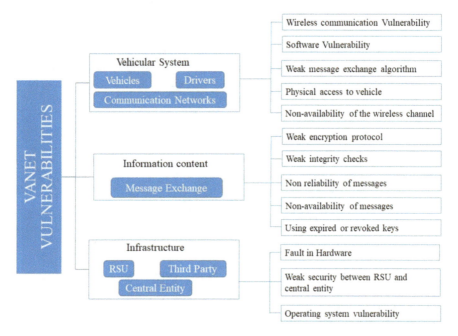

Fig. 3.15 Vulnerability taxonomy in VANET

messages might be modified by the malicious attackers resulting in alteration or non-availability of the message. This vulnerability in information exchange might have negative impact on the system.

Some of the major attacks in the vehicular system are jamming attacks, sensor impersonation attack, bogus information attack, remote access attack, physical attack, Malware attack, social engineering attack, privacy attack, eavesdropping attack, man-in-the-middle attack and wormhole attacks. The major attacks on information are eavesdropping, jamming attacks, impersonation attacks, MITM attacks, and spoofing attacks. Similarly, attacks on infrastructure are Sybil attacks and false message between central entity and RSU.

3.5.7.3 Privacy in VANET

In VANET communication, privacy and authentication are of utmost importance to provide full security to the vehicular nodes. The system verifies the identity of the vehicle user while providing privacy to the user's private data, i.e., his private and confidential information will not be revealed. However, the user should be traceable by the authority in case of legal issues like accidents. Some of the privacy and authentication are similar with the one mentioned in Sect. 3.3. We will summarize some more privacy and authentication requirements for VANETs.

- *Anonymity*: Anonymity of the individual vehicular nodes should be maintained while communicating with other nodes or infrastructure so that his real identity is not disclosed as well as privacy is maintained. To preserve the identity privacy, the nodes should send the data anonymously so that his identity cannot be traced but in case of his malicious activity, his identity cannot be detected. Therefore, there is a trade-off between privacy and traceability.
- *Anonymous authentication*: The anonymous authentication protects unauthorized vehicular nodes from gaining access to the system and prevents from unauthorized attackers. We need to consider the privacy while designing anonymous authentication for the VANET nodes.
- *Unlinkability*: The real identity of the vehicular nodes should not be linked with his location or other parameters. The vehicle should use a temporary identity known as pseudo id, which is heterogeneous to the real identity of the vehicle. The temporary identity of the vehicle should change frequently overtime to achieve unlinkability. Even if the attacker tries to obtain his pseudo id, the attacker cannot relate them to the real identity that protects the vehicle from tracking.

Figure 3.16 shows the privacy based on authentication in VANET. It can be classified into three categories. They are identity-based authentication scheme, group signature-based, and cryptography-based scheme. We describe the three categories briefly below.

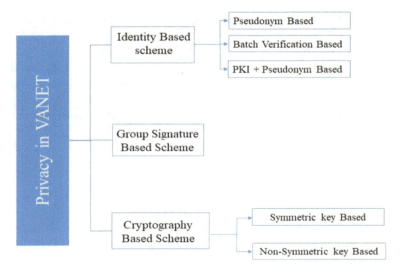

Fig. 3.16 Privacy in VANET

1. *Identity-based authentication scheme*: The identity-based authentication scheme is based on cryptography and authentication that simplify the certificate management process by using vehicle's identity while signing and verification of digital signatures. It reduces the communication overhead. It is of three types and they are as follows:
   - *Pseudo-ID based*: This scheme uses ID-based signature verification as well as self-generated pseudonyms with specifications of V2I and V2V authentication [20]. Likewise, dual pseudonymous authentication scheme has been presented earlier [21]. The vehicle receives first pseudonym from the trusted authority (TA), and it changes the pseudonym after certain period through RSU.
   - *Batch verification based*: In case of threat detection, the identity-based batch verification method for vehicle to RSU communication scheme provides the user privacy information to the system that can be traced back to the malicious user [22].
   - *PKI and pseudonym based*: A software-based solution was used for shared secret keys to maintain the privacy of the users. In addition, it has lower message overhead as compared to the previous versions [23].
2. *Group signature based*: In a group signature-based scheme, the group signature ensures the source anonymity as well as the integrity of the information. Each member of the group anonymously signs the message before sending and can be verified easily. In addition, the identity of the each vehicle that signs the message can be revealed only by the manager of the group while maintaining the privacy of the vehicles [24].
3. *Cryptography-based scheme:* This scheme utilizes the cryptography on the privacy-preserving authentication to preserve the vehicle anonymity, reduce the

communication overhead, non-repudiation, and prevent from misbehaving vehicles. This scheme is of two types, i.e., symmetric key and asymmetric key based. In case of symmetric key-based scheme, they used the symmetric keys for message authentication [25]. Each vehicle uses shared group keys or own key for creating and verifying the message authentication codes (MAC). In case of asymmetric key-based scheme, public and private keys are used for group signing and creating digital signatures. It needs to connect to the RSU for verifying the certificates, to revoke keys for malicious vehicles or collect keys from new vehicles that are registered to the RSU [26]

3.5.8 Trust Management in VANET

Besides, security and privacy in VANET, researchers worked on several trust management system to ensure the message safety. In VANET, message safety is very crucial as the vehicles communicate with neighbor vehicles and infrastructure frequently. If the attacker changes the message content such as insert bogus message or falsified message, then it might cause serious damage to the network as the false message force the legitimate vehicles to take unsafe actions resulting in fatal accidents. An efficient trust management system should be developed that can prevent the message exchange in vehicular communication from malicious attackers in real time. It should be decentralized, scalable, robust, and maintain the vehicular node's privacy. According to [27], the trust management system can be categorized into infrastructure-based, non-infrastructure-based, PKI-based, and non-PKI-based system as shown in Fig. 3.17.

The trust management schemes based on infrastructure use RSU or central authority for trust evaluation but without depending upon PKI such as TRIP [24] and RaBTM [25]. On the other hand, the trust management system based on infrastructure and PKI can effectively identify attacker with some degree of accuracy such as DTE [28], ESA [29], etc. There are researchers who work on solving the issues in message trustworthiness based on PKI but without depending upon the infrastructures [30]. The existing system has limitations such as single point of failure in infrastructure-based environment and consumes high computation capacity while using PKI and cryptography. Some researcher proposed trust management scheme

Fig. 3.17 Classification of trust management system

that works in a distributed environment in an infrastructureless environment without using PKI that helps to overcome the limitation of the other trust management system [31, 27].

3.5.9 Blockchain as a Security in VANET

Blockchain is a distributed, decentralized immutable ledger that works in trustless environment. Some of the features of blockchain are anonymity, decentralization, transparency, immutability, traceability, and non-repudiation. Because of these features, blockchain has gained lots of attention from different sectors such as industry, academia, health, and financial sectors. Some of the popular blockchain technologies are bitcoin, ethereum, litecoin, etc. As the vehicle is becoming intelligent and autonomous, they exchange data for communicating with different sensors as well as peer vehicles and infrastructures in intelligent transportation system. The message exchange between the vehicles is vulnerable to different types of attacks as the vehicles do not trust each other. The blockchain can be a solution to secure the communication between untrusted vehicular nodes in VANET. The blockchain is considered to tolerate the untrustworthy vehicles within the network because of the use of consensus protocol in the trustless environment. There are three major types of blockchain as shown in Fig. 3.18.

1. *Public blockchain*: The public blockchain is a global blockchain where any nodes can join the network, read, write, and participate in the consensus mechanism such as proof of work (PoW), and proof of stake (PoS) to determine if new block is added to the blockchain and know the current state of the blockchain. It is fully decentralized and secure as all the participating nodes verify the transactions in the blockchain. However, it is inefficient and consumes power as well as time because all the participating nodes should verify the transactions.
2. *Private blockchain*: The private blockchain is a blockchain which is controlled by a single highly trusted owner of the blockchain. It is partially decentralized network. The nodes need permission to participate and access the blockchain. The nodes require authorization to read and write the information onto the

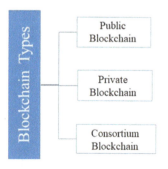

Fig. 3.18 Types of blockchain in VANET

blockchain and public auditability can be restricted. This type of blockchain is efficient because the blockchain owner does the verification. However, the single controlling power might lead to 51% attack in the network.
3. *Consortium blockchain*: The consortium blockchain is a kind of hybrid blockchain, which is the combination of public and private blockchain. It is also partially decentralized network. Only few predetermined nodes control the consensus mechanism. They have authority to read, write, and verify the blocks in the blockchain. The selection of predetermined nodes varies for different entity on the blockchain. The advantage lies in between the public and private blockchain where it takes relatively lower time to verify the transactions by the predetermined nodes.

In VANET, although several security mechanisms have been proposed to solve the security issues, the privacy and trust management issues are still of major concern. Other security mechanisms are successful in securing the vehicular communication from external attacks but not the internal attacks as the forged messages are difficult to solve. There are some attacks that analyzes the broadcasted messages between the vehicles and forge or modify the message that cause severe effects such as deroute the vehicle from its original destination and event serious accidents. Thus, the blockchain can be used to store the event logs as transactions including vehicle node trustworthiness and message trustworthiness in the blocks of blockchain that acts as a ground truth for forensics when there is a serious accident [32]. Several researchers have used blockchain technologies in vehicular networks based on cryptocurrencies such as bitcoin [33] and ethereum [34] while other have modified or proposed a new type of blockchain [32] suitable for vehicular networks. Similarly, overlay network based on block chain technology has been proposed in automotive security [10]. The overlay networks use some nodes that can manage blocks called overlay block managers in the blockchain. The nodes in overlay network have cluster heads that operates and manages the blockchain. Rowan et al. proposed blockchain for securing intelligent vehicles communication based on acoustic channel and visible light communication [35]. They have used different types of communication for securing the intelligent vehicle that is based on public blockchain and side channels using session cryptographic keys. Alternative distributed ledger systems have developed with entirely different methods of consensus mechanisms, such as directed acyclic graphs (DAGs), which no longer require the formation of a block chain, but rather use alternative cryptoeconomic methods to achieve consensus. For instance, projects such as "IOTA," "Byteball," or "Nano." The authors in [36] proposed IOTA-VPKI, a vehicular PKI (VPKI) based on IOTA distributed ledger technology (DLT) that enhances the state of the art and eliminates a single point of failure with seamless user scalability. Several works have been done in intelligent and autonomous vehicle using blockchain and details can be found in Chap. 6.

3.6 Summary

In this chapter, we discuss about the automation, intelligence, and connectivity in IAV and their implementation concerning safety, security, and privacy. The malicious attackers use different means of attack strategies at different levels, and it severely affects the vehicle's physical and cyber security system. The malicious nodes can tamper the vehicle sensors such as onboard systems, in-vehicle sensors and can intercept message exchange between the vehicles. The malicious nodes can be an insider attacker that may attack both the in-vehicle and inter-vehicle communication. Several attacker models have been demonstrated. It provides some of the security and privacy threats such as fake information attack, message replay attack, integrity, non-repudiation, access control, and privacy attack. The vulnerability is the weak point in the autonomous vehicle system that is misused and easily attacked by the attackers for their own advantages. A detailed autonomous vulnerability taxonomy of the vehicles is given, and solution mechanisms such as preventive, active, and passive defense are also provided. It also deals with the privacy of the autonomous vehicles. Some of the privacy measures discussed are cryptography-based schemes, trust management schemes, and blockchain schemes.

References

1. M.P. Babitha, K.R.R. Babu, Secure cloud storage using AES encryption, in *2016 International Conference on Automatic Control and Dynamic Optimization Techniques (ICACDOT)* (2016), pp. 859–864
2. L. Batina, J. Lano, N. Mentens, S.B. Ors, B. Preneel, I. Verbauwhede, Energy, performance, area versus security trade-offs for stream ciphers, in *The State of the Art of Stream Ciphers: Workshop Record, Brugge* (2004), pp. 302–310
3. M. Bafandehkar, S.M. Yasin, R. Mahmod, Z.M. Hanapi, Comparison of ECC and RSA algorithm in resource constrained devices, in *2013 International Conference on IT Convergence and Security (ICITCS)* (2013), pp. 1–3
4. C. Gentry, A Fully Homomorphic Encryption Scheme (Stanford University, 2009)
5. A. Acar, H. Aksu, A.S. Uluagac, M. Conti, A Survey on homomorphic encryption schemes. ACM Comput. Surv. **51**(4), 1–35 (2018)
6. G. De La Torre, P. Rad, K.K.R. Choo, Driverless vehicle security: Challenges and future research opportunities. Future Gener. Comput. Syst. (2018)
7. NIST, Framework for Improving Critical Infrastructure Cybersecurity V1.1 (2016)
8. P. Bedi, V. Gandotra, A. Singhal, H. Narang, S. Sharma, Threat-oriented security framework in risk management using multiagent system. Softw. Pract. Exp. **43**(9), 1013–1038 (2013)
9. N. Shevchenko, B.R. Frye, C. Woody, Threat Modeling for Cyber-Physical System-of-Systems: Methods Evaluation (September, 2018)
10. Microsoft, Threat Modeling for Drivers. [Online]. Available: https://docs.microsoft.com/en-us/windows-hardware/drivers/driversecurity/threat-modeling-for-drivers [Accessed: 10-Mar-2020] (2018)
11. A. Karahasanovic, P. Kleberger, M. Almgren, Chalmers Publication Library Adapting Threat Modeling Methods for the Automotive Industry Adapting Threat Modeling Methods for the Automotive Industry (2017), pp. 1–10

References

12. W. Xiong, F. Krantz, R. Lagerström, Threat modeling and attack simulations of connected vehicles: A research outlook," ICISSP 2019 Proc. 5th Int. Conf. Inf. Syst. Secur. Priv. 479–486 (2019)
13. D.P.F. Möller, R.E. Haas, *Guide to Automotive Connectivity and Cybersecurity* (2019)
14. C. Ebert, E. Metzker, Functional safety and cyber-security—experiences and trends. Funct. Saf. Symp. 1–25 (2018)
15. A. Greenberg, Hackers remotely kill a jeep on the highway—with me in it. Wired [Online]. Available: https://www.wired.com/2015/07/hackers-remotely-kill-jeep-highway/. [Accessed: 23-Aug-2019] (2015)
16. V.L.L. Thing, J. Wu, Autonomous vehicle security: A taxonomy of attacks and defences, in *2016 IEEE International Conference on Internet of Things (iThings) and IEEE Green Computing and Communications (GreenCom) and IEEE Cyber, Physical and Social Computing (CPSCom) and IEEE Smart Data (SmartData)* (2016), pp. 164–170
17. S. Jadhav, D. Kshirsagar, A survey on security in automotive networks, in *2018 Fourth International Conference on Computing Communication Control and Automation (ICCUBEA)* (2018), pp. 1–6
18. K. Mawonde, B. Isong, F. Lugayizi, and A. M. Abu-Mahfouz, "A Survey on Vehicle Security Systems: Approaches and Technologies," in *IECON 2018—44th Annual Conference of the IEEE Industrial Electronics Society* (2018), pp. 4633–4638
19. M.R. Moore, R.A. Bridges, F.L. Combs, M.S. Starr, S.J. Prowell, Modeling inter-signal arrival times for accurate detection of CAN bus signal injection attacks: A data-driven approach to in-vehicle intrusion detection, in *Proceedings of the 12th Annual Conference on Cyber and Information Security Research (CISRC '17)* (2017), pp. 1–4
20. C. Sun, J. Liu, X. Xu, J. Ma, A privacy-preserving mutual authentication resisting DoS attacks in VANETs. IEEE Access **5**, 24012–24022 (2017)
21. U. Rajput, F. Abbas, H. Eun, R. Hussain, H. Oh, A two level privacy preserving pseudonymous authentication protocol for VANET, in *2015 IEEE 11th International Conference on Wireless and Mobile Computing, Networking and Communications (WiMob)* (2015), pp. 643–650
22. C. Zhang, P.-H. Ho, J. Tapolcai, On batch verification with group testing for vehicular communications. Wirel. Netw. **17**(8), 1851–1865 (2011)
23. M. Raya, J. Hubaux, Securing Vehicular Ad Hoc Networks, vol. 15 (2007), pp. 39–68
24. J. Guo, J.P. Baugh, S. Wang, J. Guo, J.P. Baugh, S. Wang, A Group Signature Based Secure and Privacy- Preserving Vehicular Communication Framework, (May, 2007)
25. X. Lin, X. Sun, X. Wang, C. Zhang, P. Ho, X. Shen, TSVC: Timed efficient and secure vehicular communications with privacy preserving. IEEE Trans. Wirel. Commun. **7**(12), 4987–4998 (2008)
26. X. Lin, S. Member, X. Sun, P. Ho, GSIS : A Secure and Privacy-Preserving Protocol for Vehicular Communications, vol. 56, no. 6 (2007), pp. 3442–3456
27. R. Shrestha, S.Y. Nam, Trustworthy event-information dissemination in vehicular Ad Hoc networks, Mob. Inf. Syst. vol. 2017 (2017)
28. M. Raya, P. Papadimitratos, V.D. Gligor, J. Hubaux, On Data-Centric Trust Establishment in Ephemeral Ad Hoc Networks (2007)
29. M. Raya, A. Aziz, J. Hubaux, Efficient Secure Aggregation in VANETs (2006), pp. 67–75
30. Z. Liu, J. Ma, Z. Jiang, H. Zhu, Y. Miao, LSOT: A lightweight self-organized trust model in VANETs. Mob. Inf. Syst. **2016**, 18–22 (2016)
31. D. Florian, L. Fischer, P. Magiera, VARS : A Vehicle Ad-Hoc Network Reputation System (2005), pp. 0–2
32. D. Communications, A New-Type of Blockchain for Secure Message Exchange in VANET
33. Y. Park, C. Sur, K.-H. Rhee, A secure incentive scheme for vehicular delay tolerant networks using cryptocurrency. Secur. Commun. Netw. **2018**, 5932183 (2018)
34. B. Leiding, P. Memarmoshrefi, D. Hogrefe, Self-managed and blockchain-based vehicular ad-hoc networks, *Proc. 2016 ACM Int. Jt. Conf. Pervasive Ubiquitous Comput. Adjun.—UbiComp '16*, no. January (2016), pp. 137–140

35. S. Rowan, M. Clear, M. Gerla, M. Huggard, C. Mc Goldrick, Securing vehicle to vehicle communications using blockchain through visible light and acoustic side-channels, eprint arXiv:1704.02553
36. A. Tesei, L. Di Mauro, M. Falcitelli, S. Noto, P. Pagano, IOTA-VPKI: A DLT-based and resource efficient vehicular public key infrastructure, in *2018 IEEE 88th Vehicular Technology Conference (VTC-Fall)* (2018), pp. 1–6

Chapter 4
In-Vehicle Communication and Cyber Security

4.1 Overview

The evolution of intelligent and autonomous vehicles increases the dependence on information sharing among in-vehicle electronic control units (ECUs) and communication within the vehicle, and increases the connection with other vehicles. As the category of safety equipment is concerned, driver assistance systems for safe driving are experiencing a dramatic growth and development process. Modern revolutionary automatic driving assistance system (ADAS) functions are made up of dynamic integrated and networked cyber-physical structures. While the connectivity to the outside world allows many new services, it also exposes the vehicle and its electronics to a possible remote attack. Hackers with experience of reverse engineering can access the vehicle's electronic components quickly and in a very short time, resulting in the opening of a new attack surface.

We discuss the different types of in-vehicle networking (IVN) systems and their security threats. In Sect. 4.2.1, we discussed seven categories of vehicle electrical and electronics (VEE) used in the in-vehicle system. The vehicles use ECUs to communicate with other control units, sharing vital vehicle information via the LAN protocol. We also discuss the different types of IVN protocols such as CAN, FlexRay, automotive Ethernet, LIN, and MOST along with their security threats. In Sects. 4.4 and 4.5, we present the IVN architecture and its challenges on OBD-II ports, threats, and countermeasures. We then discussed the cybersecurity in IVN and presented the cybersecurity protection layer for the in-vehicle systems.

4.2 In-Vehicle System

Over recent years, vehicles become more advanced and provide new and diverse functions and facilities. To cope with the demands and requirements of passengers and the drivers, several types of electronic devices and sensors are installed in the vehicle. Currently, the number of such electronic devices for the safety of vehicles to infotainment has increased rapidly and continuously. Advanced navigation systems and multimedia devices have made their way as mandatory vehicle components so as to make driving more convenient. As a result, vehicles are increasingly evolving in the form of smart cars or connected cars in simple mechanical systems, and fully self-driving vehicles are expected to deploy commercially soon. Various electronic control units (ECUs) are installed and communicate with other control units, exchanging critical vehicle-related information through in-vehicle LAN communication protocol. More than 100 embedded ECUs could be integrated in advanced vehicles. The evolution of intelligent and autonomous vehicles increases the dependence on information sharing among in-vehicle ECUs and communication within the vehicle, and increases the connection with other vehicles, resulting in the opening of a new attack surface. The success of vehicle attack is based on three basic classes: remote attack surface, cyber physical features, and in-vehicle network architecture. To defend against this, a strong security solution is required in some form.

4.2.1 Vehicle Electrical and Electronic System

Vehicle electrical and electronic (VEE) covers all electrical as well as electronic systems or components used in the vehicle. Regarding the category of safety devices, the vehicle assistance systems for safe driving are undergoing a rapid growth and expansion process as they shift from high-end to the mid-range cars. The VEE components can be divided into seven categories as shown in Fig. 4.1 [1]. We will briefly describe each of these categories.

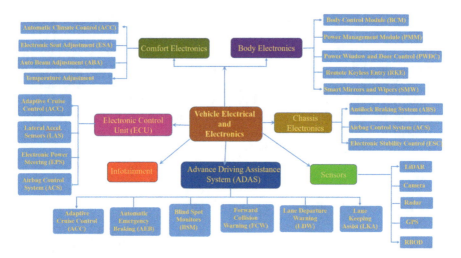

Fig. 4.1 In-vehicle electrical and electronic system

4.2 In-Vehicle System

4.2.1.1. Body electronics: Functional hardware for body electronics is very important for vehicle owners because they are searching for new comfort levels, performance, safety as well as other features in their vehicles. Some of the body electronics are:

   a. Body control module (BCM) controls and monitors different vehicle electronic parts and communicates with ECU via bus systems like CAN and LIN. BCM controls several vehicle parts such as air conditioner, power windows, central locking systems, and power mirrors.
   b. Power management module (PMM) manages the power and provides power to VEE (like ECU, motors, lights, etc.) as well as monitors and controls devices.
   c. Power window and door control (PWDC) is responsible for handling and controlling VEE via CAN and LIN bus. It controls door lock, mirror axis control, light bulbs, windows, etc.
   d. Remote keyless entry (RKE) is the vehicle electronic remote locking system that controls vehicle accessibility. It triggers without physical contact via a remote control device that works as standard vehicle key. RKE can be used to remotely lock/unlock doors, start the engine, or auto park.
   e. Smart mirrors and wipers (SMW) are activated automatically when the sensors sense that it is raining. It adjusts the wiper speed according to the intensity of the rain.

4.2.1.2. Chassis Electronics: The chassis electronic system contains several components that tracks, monitors, and actively controls various parameters as shown in Fig. 4.1. Some of the chassis electronics are:

   a. Antilock braking system (ABS) allows the car's tires to maintain traction on the road surface based on the driver's inputs while braking, to keep wheels from locking up, and to avoid uncontrolled slips on the slippery road surfaces.
   b. Air bag control system (ACS) protects the driver and passengers during a crash event to avoid any major forces during the accident. It saves driver, passengers, and the interior of vehicle.
   c. Electronic stability control (ESC) increases the stability of the vehicles by sensing and reducing traction loss. Automatic braking is applied to the wheels if the vehicles lose control of the steering.

Comfort electronics are electronic devices installed in the vehicles that allow a comfortable ride for the driver and the passengers such as

   a. Automatic climate control (ACC) regulates and adjusts the temperature of the vehicle cabin with respect to the outside temperature, weather, etc.

b. Electronic seat adjustment (ESA) keeps track of each user's seating position along with the mirror adjustment, which can be retrieved to corresponding drivers' settings if the driver is changed.
c. Auto beam adjustment (ABA) adjusts the lightening conditions during morning and evening based on different types of photosensors.
d. Temperature adjustment regulates and controls the air-conditioning temperature based on automatic system considering the air temperature information collected from the temperature sensors installed in the vehicle.

4.2.1.3. Electronic Control Units (ECUs): An ECU is composed of hardware and software, where the software is a firmware based on the specific characteristics of the ECU. ECU consists of erasable programmable read-only memory (EPROM), flash memory, electronic solid-state nonvolatile storage medium, communication interfaces, protocols, and other electronic components. We will discuss more the ECU in Sect. 4.4. There are three types of ECU based on their primary functionality, and they are:

a. Engine control module (ECM) is the main module in the vehicle system. ECM runs the actuators based on the collected information from the sensors and regulates the settings. The ECM control requirement is one of the highest real-time constraints due to the use of several measuring sensors. ECM has logic circuits for performing actual controlling.
b. Transmission control module (TCM) monitors and controls the transmission system like gear shifting. The automatic transmission needs control for its operation. Most semiautomatic transmissions also have fully or semiautomatic clutches. The ECM and TCM exchange information, sensor signals, and control signals for their services.
c. Vehicle control module (VCM) is coupled with several types of sensors to control and monitor different systems in the vehicle. When there is a crash, the VCM gets signals from the accelerometers, and then it deploys the air bags based on different parameters of the sensors. The VCM takes input from various sensors to provide the electronic stability control (ESC) data for the safe operating situation.

4.2.1.4. Infotainment Electronics: The infotainment electronics are developed by third-party suppliers or OEMs. It provides in-vehicle entertainment to the passengers and the drivers while driving on the road such as audiovisual entertainment like movies or songs. The basic infotainment systems are described as follows:

a. In-Vehicle infotainment (IVI) such as audio systems (radio/CD players) and video players that play videos on the tablet screens inside the vehicles and navigation systems are based on in-vehicle Internet, Bluetooth, USB, GPS, Wi-Fi and cellular systems.

4.2 In-Vehicle System

b. Navigation systems display the geographical location of the vehicles on the map via GPS sensors on a display unit, suggest shortest and best destination, traffic information, etc.

c. Vehicle audio systems provide the music entertainment system and voice navigation system inside the vehicle that can be operated from dashboard and voice command or controlled by the controller located on the steering.

Sensor technology used in the autonomous vehicles is of different types such as electric, mechanical, optical, sound, image, and light sensors. The sensors used in the vehicles should have high accuracy, high precision, high resolution, high sensitivity, and less noise and should consume less power. Some examples of sensors are

a. Light detection and ranging (LiDAR) consists of GPS, scanners, and laser technologies to generate 3D information of a particular area providing remote sensing based on pulses of light. It uses both position and velocity to locate and identify the objects around the vehicles. Although LiDAR-based devices can do exceptionally well in some range measurement scenarios, they are unable to obtain visible traffic details such as traffic signs, traffic signals, on-road text, and color or shades on the ground.

b. Cameras used in automotive vehicles are exterior infrared cameras known as night vision cameras or interior cameras, also known as gesture tracking cameras for monitoring the driver's behavior and gestures like drowsy drivers, eyelids closed or sleepy pupils, face recognition, etc. The Tesla Model 3 uses eight exterior cameras for capturing 360° views.

c. Radio detection and ranging (RADARs) are sensors that are robust to rain or snow and are sensitive to installation tolerances and materials. The range for RADAR is more than 200 m for headlong direction and 50–100 m in all other directions.

d. Sensor fusion combines the information from different sensor technologies and makes appropriate decision like human drivers in all circumstances. Moreover, it should be such that the technology should be easy to use and people should embrace the technology. The aim of the sensor fusion is to evaluate the situation around the direction of the vehicle with enough precision, confidence, and low latency to safely navigate the vehicle.

4.2.1.5. Advanced Driver-Assistance Systems (ADAS): ADAS is intelligent driving system that provides important information related to road traffic, congestion levels, emergency warning, route suggestion, lane change, emergency braking, etc. The system is installed inside the vehicle, and it is designed to help drivers to lower their workload while driving and parking [2]. It alerts drivers regarding potentially hazardous conditions or takes over vehicle

Table 4.1 Types of ADAS and their functions [2]

ADAS	System function	Region
Adaptive cruise control (ACC)	Dynamically changes vehicle speed to maintain a fixed trailing gap from a leading car, assisting drivers to continually brake	Forward
Automatic emergency braking (AEB)	Detect slow traffic or stopped vehicle ahead and apply the brakes immediately	Forward
Blind spot monitoring (BSM)	Warns drivers to the presence of cars on each side at the blind spot by showing an indicator in the corresponding door mirror	Lateral
Forward collision warning (FCW)	Notify driver of a moving or stationary car in front; prevents from a potential collision	Forward
Lane departure warning (LDW)	Use visual, vibration, or sound alerts; lane exit warning system notification if driver moves off the path	Lateral
Lane keeping assist (LKA)	Helps driver to return to his lane if he drifts out of the lane based on the lane markers	Lateral

control in hazardous situations. These systems can also be used to assess the human driver's exhaustion and stress level, providing preventative warnings or assessing driving performance and making suggestions about it. Such systems can switch over human controls on any threat assessment and perform simple tasks such as cruise control or complex movements such as parking and overtaking. The main advantages of using the ADAS are that they allow communication between neighbor vehicles, transportation maintenance services, and traffic management centers. Some of the advanced driver-assistance systems are given in Table 4.1 and are described below:

a. Adaptive cruise control (ACC) dynamically changes vehicle speed to maintain a fixed trailing gap from a leading vehicle, alleviating drivers from needing to continually brake and reset the cruise control (or monitor speed and progress made), when a vehicle ahead of them decelerates. ACC has demonstrated significant safety advantages in the context of improved vehicle movement and less forward warnings of collisions.

b. Automated emergency braking (AEB) technology will reduce or, in certain situations, avoid rear end and other spatial collisions. This function will detect slow traffic or detect stopped vehicle ahead and apply the brakes immediately if the driver fails to respond.

c. Blind spot monitoring (BSM) employs rear bumper-mounted RADARs or other sensors to identify automobiles coming from behind and oncoming lanes. This warns the driver regarding the presence of vehicles on each side at the blind spot by showing an indicator in the corresponding door mirror. When the driver tries to change the lane in the

4.2 In-Vehicle System

blind spot area, the led icon appears and it uses a symbol, tone, or vibration to warn the driver that there are vehicles in their blind spots. Its function can sense slow or stopped traffic and apply the brakes urgently if the driver fails to respond.

d. Forward collision warning (FCW) will notify drivers of a fast moving or stationary cars in front of them and prevents a potential collision. FCW utilizes a sensor to identify cars or objects in front of the car. The device measures the distance to the front target and, if the car gets close enough to the danger of accident, warning sounds and displays a visual message, urging the driver to apply the brake.

e. Lane departure warning (LDW) detects road surface line markers and alerts the driver of unintended lane departures. This system is particularly suitable during conditions when the path is consistently straight, and drivers seem not to pay enough attention on the lane. The system accepts the movement as deliberate, when the lane shift is followed by turn signals or acceleration, and does not trigger a warning.

f. Lane keeping assist (LKA) feature helps the driver to return to his lane if he drifts out of the lane based on the lane markers on the road and prevent from a serious accident. LKA triggers LKA warning alerts, and the driver sees a message on the dashboard, hears a sound, or feels vibration on his/her seat. If the driver does not react in time, one will be guided by LKA, returning the vehicle to the center of the path.

4.3 In-Vehicle Communication

The in-vehicle communication refers to the intra vehicle communication where all the internal components like telematics, sensors, and actuators communicate with each other using different communication mediums like standard bus system. The in-vehicle networks are also called intra-vehicle networks or automotive networks. The in-vehicle communication occurs within the internal vehicle communication network where the ECUs intercommunicate with other electronic subsystems. The ECUs are connected within vehicle bus systems through specialized internal communication networks. Most ECUs are connected to one or more bus networks for monitoring and controlling the vehicles. Some ECUs are connected to external controls, such as digital equipment, infotainment, and navigation systems via a gateway system. The central gateway-based architecture links the entire IVN with a central gateway and provides smooth connectivity between heterogeneous network protocols.

There are different vehicle domains that have specific requirements, which led to the development of a large number of automotive bus network such as local interconnect network (LIN), control area network (CAN), FlexRay, Media Oriented Systems Transport (MOST), automotive Ethernet, and low-voltage differential signaling (LVDS) [3]. We will discuss more the vehicle domains in Sect. 4.3.2.

4.3.1 In-Vehicle Sensing Technologies

To achieve full self-driving intelligence, the vehicles need to observe their own condition, surrounding environment, and other situations beyond their visual range. Figure 4.2 shows the gateway block diagram that interconnects different components of in-vehicle system with the gateway. The perception of vehicle self-state and decision making are based on the installed high-precision sensitive sensors such as pressure, engine temperature, and speed sensors. The driving safety can be ensured by using the sensors installed in the vehicle that captures perception of the surrounding environment of the vehicle such as obstacles on the road, driving condition, and environmental condition. In case of high-speed driving, the safety of the vehicle can be ensured by long-range sensors or other communication modes. Based on this information obtained from sensors, the vehicle will accomplish the self-driving control, independently based on environment sensing, decision making, and other driving controls. The in-vehicle technologies can be further categorized into three different technologies.

43.1.1. Sensor Technologies: It includes LiDAR, VLS, ultrasonic ranging device (URD), infrared ranging, and millimeter wave radar (MWR). The LiDAR consists of a GPS, scanner, and laser technology to generate 3D information about a particular area, providing remote sensing based on pulses of light; for example, multiple LiDAR composed of 64-detector array are used in vehicles.

4.3.1.2. Vision Technologies: It includes stereo vision system (SVS), HD cameras, black box, or CCTVs. They provide visual information for analyzing critical situations such as road accidents. It helps in forensics and takes necessary actions by recording visual information with high confidentiality, authenticity, and integrity. SVS reconstructs a 3D video using multiple viewpoints and structured light source. There are two methods used to construct a 3D video; they are active and passive method. The active method is based on

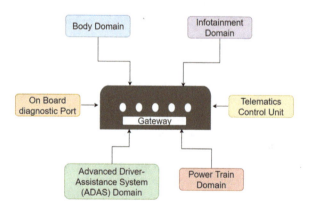

Fig. 4.2 Gateway block diagram

4.3 In-Vehicle Communication

complementary structured light source, while the passive method is based on unstructured light sources captured by the camera.

4.3.1.3. Positioning Technologies: Some of the positioning technologies are GPS, radar cruise control, and radar-based obstacle detection (RBOD). The GPS receiver can be used along with the Doppler radar speedometers, RCC, and RBOD in autonomous vehicle to provide precise location of the vehicles and active location validation.

4.3.2 In-Vehicle Network (IVN) Systems

Due to the requirement of advanced vehicles such as electric propulsion, hybrid, or driverless vehicles, they need more stringent real-time information and to fulfill this requirement, the application of CAN in vehicles is evolving rapidly. There are other in-vehicle network (IVN) systems for complex and heterogeneous architectures such as FlexRay, automotive Ethernet networks, local interconnect network, and Media Oriented Systems Transport (MOST). Since a large number of sensors are used in automotive vehicles, a big amount of data is produced every second. In addition, all those sensors need to communicate with each other with strict minimum latency. Different components in the vehicles require a different bandwidth and latency and continue to increase in complexity along with the increase in the number of ECUs used in automotive vehicles. The details of in-vehicle networking protocols are discussed as follows and are shown in Fig. 4.3.

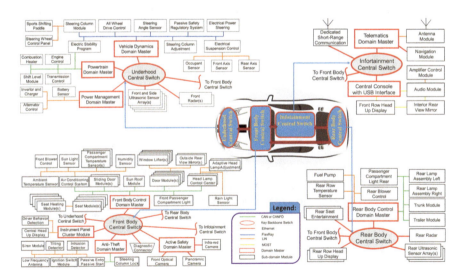

Fig. 4.3 In-vehicle networking system [4]

4.3.2.1. Controller Area Network (CAN): CAN was introduced in 1986 by Bosch focusing on reliability and functional safety of the in-vehicle communication network. The CAN bus system is the dominant standard used by automotive industry for in-vehicle networking and communication to transmit information among the different sensor technologies. The reason for its wider adoption is due to its cost efficiency and flexibility compared with other network technologies. CAN is used for connecting electronic vehicle parts, such as microcontrollers and other devices, to communicate without a host computing machine. It has other features like resistance and fault tolerance against external interference by using unshielded twisted pair lines. It has capability to detect as well as report errors to physical layer. However, there are some issues with CAN bus system like Byzantine Generals Problem (BGP) and incapable of handling bubbling failure. A solution has been proposed to prevent local and remote attacks by enabling ECU with authentication, integrity check, and frame encryption while transmitting [5]. There is an alternative of CAN known as CAN with flexible data-rate (CAN-FD) that uses different data rates during one cycle of information transmission. It sends higher data rate while transmitting actual bits and normal data rate while transmitting arbitration bits [6]. The CAN payloads permit 8 bytes (B) of data, whereas CAN-FD permits up to 64 B of data. Additionally, by fixing the bit rate switch, CAN-FD allows speeds of up to 8 Mbps as compared to 1 Mbps of CAN. Vector has developed development and testing software tool called CANoe that has the ability to run fuzzy testing effectively and efficiently for individual ECUs and entire ECU networks in the automotive industry. It supports CAN, MOST, LIN, FlexRay, AE bus system along with CAN-based protocols like J1939 [7]. It offers support during the entire development process, from design to system level testing to network designers, developers, and test engineers. OEMs and suppliers globally effectively leverage their flexible features and configuration solutions.

4.3.2.2. FlexRay: The consortium of major companies from the automotive industry and leading suppliers developed a standard automotive networking called FlexRay. FlexRay is resilient to errors related to communication channel, sampling signals, redundancy check, coding, and decoding. FlexRay interaction exists in the form of connection cycle and can include static as well as dynamic segments. Static segments are based on time-division multiple access (TDMA) for real-time communications, whereas the dynamic segment is based on an event-driven communication protocol for service-based messaging. The FlexRay bus is a fault tolerant and flexible bus system providing higher data rate at higher speed than CAN. The FlexRay has data rate of 10Mbps as opposed to CAN, which has rate of 1Mbps. However, FlexRay is unfavorable to be deployed in automobiles as compared to CAN and LIN due to its higher cost and complex communication protocol, but it is used in safety-critical applications in which sent information has to meet a strict time frame [8]. For industrial case, a divide-and-conquer method

4.3 In-Vehicle Communication

was proposed to solve the subproblems in cascade mode based on TESLA protocol [9].

4.3.2.3. Local Interconnect Network (LIN): In November 2002, the first fully implemented LIN specification was released. The new LIN technology puts together LIN's low-cost, flexibility, and basic sensors to build small networks. Such subsystems can be linked to in-vehicles through a backbone network, such as CAN. The LIN is also widely used in the automotive industry due to its lower cost because of the use of the single-wire network, flexible wire harness, and easy software development as compared to CAN. It uses *UART* ports, making a variety of microcontroller capabilities such as 8-bit microcontrollers used as LIN controllers without special requirements. The LIN is based on linear bus topology working on master–slave mode. It is robust against transmission errors due to use of checksum, parity bit, and response error bit. However, it has one drawback; i.e., the maximum transmission speed is about 20 kb/s, which is very low compared to other bus networks.

4.3.2.4. Automotive Ethernet (AE): The AE is an emerging technology that is capable of delivering a higher bandwidth of 100Mbps through its supported multi-access full duplex transmission suitable for multimedia and ADAS. The AE has better security features and low latency compared to CAN and LIN. It employs IP-based routing scheme that prevents the ECU from being hacked and prevents attackers from full control to the whole Ethernet. In case of CAN and LIN, they did not consider the security aspect thoroughly [10]. It has other features such as it requires only less number of ECUs and cables that provide high bandwidth. The AE offers a high-speed communication for in-vehicle communication. The IEEE 802.3bw standardized a gigabyte solution for automotive Ethernet in 100BASE-T1 [11]. It supports 100 Mbps speed that increases the data throughput satisfying the requirement of automotive vehicle standards as well as reducing the internal cabling weight and cost in automotive vehicle networking. Ethernet Audio Video Bridging (AVB) is a real-time audio and video Ethernet interface standard. The first version, AVB ver. 1.0, is planned to be implemented in the fields of vehicle multimedia and video capturing. The second version, i.e., AVB2.0, will fulfill vehicle inspection specifications.

The Ethernet is based on two protocols that tends to be the most common automotive applications, and they are Time-Triggered Ethernet (TTEthernet) and Time-Sensitive Networking (TSN). The TTEthernet accepts two different types of network traffic in parallel to conventional network traffic: time-triggered (TT) that includes clock synchronization, which enables data transfer to occur at precise intervals, while rate-constrained (RC) that has a lower capacity than TT data, but provides predetermined bandwidth. On the other hand, Ethernet TSN includes a range of technologies, including time synchronization as well as traffic routing, frame preemption, and ingress control, and is a high-speed networking system that is adequate to enable autonomous vehicle applications. It is not limited to serving just multimedia data categories. Due to the use of common Ethernet standards with only slight modification, its availability helps it for easy adoption in the automotive industry. Nonetheless, gigabit Ethernet is not suitable for automotive vehicles due to its higher cost.

4.3.2.5. Media Oriented Systems Transport (MOST): The introduction of MOST bus is to meet the standard of infotainment application with higher data rate inside and outside of automotive vehicles. CAN, LIN, and FlexRay are primarily used for control systems, while MOST is used for telematics applications. MOST bus systems are mainly used for infotainment purposes, enable audio, video, voice, and control data to be transmitted. The MOST is based on daisy chain or ring topology and synchronous data communication to transport multimedia and data signals via plastic optical fiber (POF) (i.e., MOST25 and MOST150) or electrical conductor (i.e., MOST 50 and MOST150) in physical layers. It is an ideal solution for ADAS as data is transmitted through optical cable. MOST transmissions include FlexRay-like static and dynamic segments, a receiver and transmitter, and a channel controller from which data channels can be asserted. Some of the features of MOST are: (i) It prevents interference as the transmitter and receivers are separated, (ii) it has high bandwidth, and (iii) it is associated with seven OSI layers. However, it has some drawbacks that using MOST in vehicle is expensive as well as a single faulty MOST node causes complete network shutdown.

4.3.2.6. Single Edge Nibble Transmission (SENT): SENT is a point-to-point scheme protocol for transmitting signal values from a sensor to a controller, and it is based on SAE J2716. It is designed to enable high-resolution data to be transmitted at a low resolution at the minimum device costs. It consists of an in-built feature, which allows the obtained data nibbles to be sorted directly in HW without any program interruption.

4.3.2.7. Other Bus Protocols: There are other similar technologies such as Byteflight (FlexRay Predecessor), vehicle area network (VAN), Time-Triggered CAN (TTCAN), and low-voltage differential signaling (LVDS). LVDS is a formal specification that defines the electrical properties of a differential communication protocol. LVDS functions at low voltage and can use inexpensive twisted pair copper cables to run extremely high speeds. It is a physical layer specification that is used by several data transmission specifications and applications. Clock eXtension Peripheral Interface (CXPI) is developed to support the commonly used vehicle networking LIN protocol. By having low speed and fewer wire harnesses in a vehicle, CXPI decreases vehicle weight and fuel consumption. It supports automotive control systems such as door control unit, light switch, and air-conditioning systems. BroadR-Reach technology is a common Ethernet physical layer intended for use in vehicle networking. The BroadR-Reach technology enables multiple in-vehicle systems to access information simultaneously through unshielded single twisted pair cable. The advantages of implementing the BroadR-Reach Ethernet standard for automakers include decreased connection costs and cabling weight. However, these technologies are struggling to be adopted widely in the automotive industry, as they need to show significant and improved features as compared to the mentioned technologies above. The detailed comparison of all the automotive in-vehicle bus protocols is given in Table 4.2 [12].

4.4 In-Vehicle Network Architecture and Topology

The growth in the number of different systems requires convergence and centralization which in turn leads to other improvements and certain technical changes within the vehicle. In order to enable a variety of features and requirements of in-vehicle functions like convenience, infotainment, safety, security, etc., the use of number of ECUs has been rising. Several IVN protocols have been proposed for connecting the ECUs. As discussed in the previous sections, CAN is a predominant protocol for IVNs, yet it cannot provide real-time performance that is necessary in critical security applications. To overcome the issue faced by CAN, FlexRay has been introduced; however, it cannot deliver adequate bandwidth (BW) for entertainment and multimedia applications. So, MOST has been developed to satisfy the infotainment requirements, but it has certain constraints when the BW requirements increased exponentially. Thus, automotive Ethernet has been developed to overcome all the existing issues; for instance, Ethernet is used for monitoring, connectivity, and infotainment, and can be used in future vehicle applications for ADASs and backbone networks. Since Ethernet, FlexRay, and CAN are used on the same vehicle at the same time, a gateway is required for seamless connection between them. The IVN architecture is changing from the central gateway-based architecture to an Ethernet backbone-based architecture [12]. Ethernet has a BW of more than 100 MB/s that is suitable for handling seamless connectivity than CAN that has BW of 1 MB/s and FlexRay that has BW of 10 MB/s. In central gateway-based architecture, a central gateway interconnects with the overall IVN that enables uninterrupted connectivity between heterogeneous network protocols, whereas in backbone-based architecture, a domain control unit (DCU) plays the role of the gateway between the backbone network and its own subnetwork. In backbone network, the number of gateways increases as it uses a higher number of DCUs. In fact, for each type of vehicle, a specific gateway needs to be applied accordingly. Thus, a gateway should be easy to construct, configure, and check using a variety of hardware and software technologies.

In 2016, Bosch came up with the road map of automotive IVN and future EE architecture [13]. Figure 4.4 shows the evolution of in-vehicle networking architecture developed by Bosch. The figure shows the distributed EE architecture, mainly encapsulated electronic structure and functional integration. In the past, modular EE architecture based on a central gateway architecture had its own ECU, where each function is represented as a box in the figure. The current architecture that is the part of the distributed EE architecture integrates each function of the modular architecture.

Table 4.2 Comparison of all the automotive in-vehicle bus protocols

Characteristics	CAN	FlexRay	LIN	Ethernet BroadR	Ethernet AVB	MOST	LVDS	SENT	CXPI
Date	1983	2005	2001	2011	2011	2001	2002	2007	2015
Standard	ISO 11898	ISO 10681	ISO 17987	IEEE 802.3BW	IEEE 802.1	–	ANSI TIA/EIA-644-A	SAE J2716	SAE J3076
Topology/architecture	Multi-master (10–30 nodes)	Multi-master (<64 nodes)	Single master (2–10 slaves)	Centralized network	Switch/gateway	Multi-master (<64 nodes)	P2P/multi-point/multi-drop	P2P sensors to controller	Master-slave
Physical layer	Differential twisted pair	Differential twist pair	Single wire	Unshielded single twisted pair		2 wires (optical/electrical)	Shielded twisted copper	Single wire	Single wire
Domain application	Chassis/powertrain	Safety-critical	Body electronics	ADAS, connected car, infotainment		Multimedia and telematics	Displays and camera	Safety-critical	Body electronics
Data rate	500 Kb/s	10 Mb/s	19.2 Kb/s	100 Mb/s	>100 Mb/s	150 Mb/s	4 Gb/s	30 Kb/s	20 Kb/s
Cost per node	Low	Very high	Very low	Very high	–	Very high	High	Very low	Very low
Attribute	– Low speed (125 kb/s)/fault tolerant – High speed (1 Mb/s)/flexible data	– X-by-wire – Reliable	Cheap	– Lower cost – Reduced weight	– Transmit and receive via gateway	– Lighter – Meet EMC requirements	– Low power – High speed – Low noise	– One-way digital precision	– Alternative to LIN for real-time HMI system

4.4 In-Vehicle Network Architecture and Topology

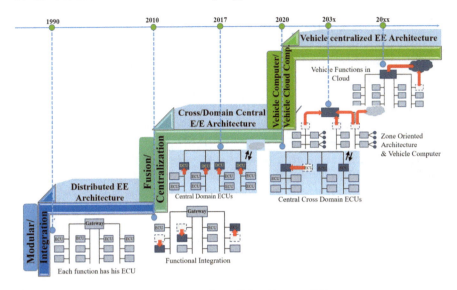

Fig. 4.4 Road map of automotive IVN and future EE architecture [13]

In the current integration architecture, it integrates the hardware and software into the existing ECUs. A centralized architecture is the central gateway-based architecture where all the IVNs connect through the central gateway. A centralized domain control unit (DCU) acts as the gateway between the Ethernet backbone networks. Ethernet is anticipated to be one of the predominant IVN protocols for future automotive vehicles in the coming years. It provides robust security compared to the current technology. Current modular architectures incorporate one feature per module. New architectures incorporate ECU functions and add more sensors. The standardization of the basic controller is under progress. After few years, cross-domain centralized EE architecture will be available to handle the complexity of increasing cross-domain functions. There will be fusion of the domains with central cross-domain ECUs.

According to Bosch [13], in the future, there will be vehicle-centralized EE architecture where there exists a zone-oriented architecture integrated with vehicle computers. This is a visionary architecture, where there will be logical centralization and physical distribution, i.e., domain-independent vehicle-centralized method with central vehicle intelligence and neural networks (acts as zones). In vehicle cloud computing, the vehicle functions will be integrated in the cloud as shown in Fig. 4.4. There will be increasing number of vehicle functions in the cloud, and each zone controls vehicle functions. This is just a prediction from Bosch, and we need to wait and watch if the vehicle-centralized EE architecture will eventually implement in the future along with the evolution of electric and autonomous vehicles.

4.5 Functional Safety and Cybersecurity

Functional safety is part of a vehicle's overall safety system or its modules that depends on the cyber physical system (CPS). It provides safety for its modules so that the modules function correctly in response to the inputs, including secure maintenance, equipment faults, etc. Functional safety standards emphasize on electrical, electronic, and programmable electronic (E/E/PE) devices and functional safety needs to be applied to non-E/E/PE elements of the network as well. In 2011, the International Organization for Standards (ISO) established ISO 26262 for the functional safety of VEE systems. The functional safety is defined by ISO 26262 and IEC 61508 for hazard and risk analysis (HARA), safety engineering, functional and risk management. The goal of functional safety is to reduce the risk to a tolerable level and to reduce its negative effects. The attacks to the functional safety result in physical damage and injuries to the drivers due to the road hazards. The attacks on the cybersecurity result in the malfunction of the operation of the vehicle components such as unable to start the vehicle and locked doors. The vehicle security is defined by security standards such as ISO 27001, ISO 15408, ISO 214534, and SAE J3061 for threat and risk analysis (TARA), security engineering, and misuse cases. Cybersecurity is the major liability risk due to the weakness in the in-vehicle security systems such as weak connectivity, open channels, potentially unsecure bus systems, and existence of the intelligent attackers that launch cyberattacks. An efficient risk analysis and management system will lead to reliable and secure vehicular networks and vice versa. By integrating the functional safety and cybersecurity based on holistic system, engineering brings reliable, efficient, and secure autonomous vehicular system.

Functional safety can be accomplished when each specified safety task is performed and the performance level needed for each safety function is met [1]. This is usually done by a method that involves steps such as (i) safety function identification, (ii) safety function assessment, (iii) safety function verification, (iv) failure mode and effects analysis (FMEA), and (v) functional safety audits activity. More details about functional safety and cybersecurity are given in Chap. 3.

One of the safety concerns in autonomous vehicle is that human-in-loop activities cause risks that can lead to failures and disasters, when appropriate safety mechanisms are not properly implemented. The functional safety and cybersecurity are interrelated with each other. The cyber-attacks by the malicious attackers on the software or communication part of the vehicle will eventually lead to the functional safety issue of the vehicles. The ISO 26262 ed.2 includes shared methods for functional safety and security such as TARA that ensures the protection of vehicles from malicious attackers. Safety and security are collaborative and involve comprehensive system engineering. Figure 4.5 shows how the combined safety and security issues are initiate for taking the appropriate steps: from design stage, to validate the same by testing and validation of the autonomous vehicles [14]. The steps are as follows:

a. Threat, risk, and hazard analysis for security and safety are carried out.
b. Safety and security goals are defined.

4.5 Functional Safety and Cybersecurity

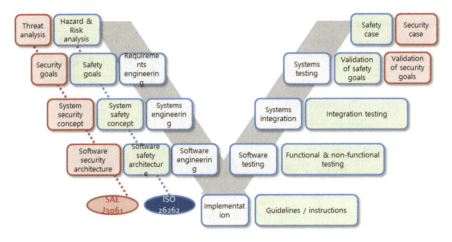

Fig. 4.5 Interaction between functional safety and cybersecurity [14]

c. Definition of security and safety principles for hardware and software modules in which safety and security roles are assigned.
d. System architecture definition.
e. System integration, validation, and testing of applications to meet security and safety criteria.
f. Instructions and documentations.

It is recommended to extend the hazard analysis with threat analysis during the designing and development phase of the autonomous vehicle to be robust against different attacks. It is better to reuse existing safety artifacts to ensure strong safety and security define tailored security protection for safety-critical systems. The entire bus communication system should be encrypted using AUTOSAR, and all the ECUs should be isolated and connected through domain control units [15].

4.6 In-Vehicle Cybersecurity Issues and Challenges

4.6.1 Challenges of IVN Architecture

While the developments regarding IVN were outlined in the previous segments, other challenges and issues remained to be solved in the IVN. Several sensors, actuators, ECUs, and microcomputers are needed to be incorporated in future vehicles. Moreover, communications with different subsystems/domains will be highly complex. This condition requires a new IVN technology with higher support for the bandwidth. From the overview of existing in-vehicle networking technology, it can be easily seen that the Ethernet, which can provide up to 1 GB/s data rate, is an optimal solution for this problem. Additionally, Ethernet's integration of various protocols

Table 4.3 Requirements of in-vehicle system

Subsystem	Description	Bandwidth requirement (Mbit)	Delay requirement (μs)	Reliability requirement
Powertrain	Calibration	>10	<10	High
	Diagnostics	>100		
	Control data	>1		
Chassis and safety	Surround view	>1000	<10	High
	ADAS	>1000		
	Control data	>1		
Infotainment	Multimedia	>10	<10	Medium
	Display	>1000		
	Diagnostics and flashing	>100		
	Connectivity	>100		

with different subnetworks/domains will lower the connectivity barrier for those networks, resulting in simpler and more effective communications between them.

However, using Ethernet for high BW and communication efficiency will bring more challenges with additional requirements for in-vehicle networks as given in Table 4.3. Some of the challenges are described below.

Bandwidth: Using Ethernet for IVN infrastructure to handle higher data speeds will only increase gateways including tools and other methodologies that will in turn increase the management effort of these technologies. Using Ethernet just to replace the existing 1:1 networks instead of constructing the system differently makes it more complicated than necessary, even with 100 Mbps of BW because it does not take full advantage of the functionality that Ethernet provides. In addition, it needs to construct like CAN over Ethernet that results in extra overhead. This implies that more extreme (but efficient) improvements cannot be done instantly. The existing networks now run simultaneously with extra costs and effort. An entire Ethernet IVN will have an effect on how we design and construct automobiles. This could shift not just the IVN but also the cycles of commodity and the value chains.

Latency: In addition to the requirement for bandwidth, latency is another important factor of in-vehicle networking. Most safety-related components and powertrains as well as chassis controls in a vehicle have stringent delay specifications in order to guarantee reliability. A number of reasons can cause latency, especially where there is connectivity through multiple domains. Imagine, for example, a transmission of data which is interrupted by other noise signals [16]. Even though automotive Ethernet offers a higher bandwidth for IVN, its multimedia connectivity poses demanding challenge of complying with safety system delay limitations and vehicle control. New automotive Ethernet technologies are also required to satisfy the criteria for vehicle latency.

Reliability: In-vehicle systems have strict specifications for control and monitoring purposes including latency and reliability for supporting the interactions of vehicle-to-sensor on board. In the near future, complete implementation of vehicle-to-sensor onboard networking will be achievable only when the vehicle-to-sensor onboard networking can deliver the same efficiency and reliability as the wired communication. The powertrain systems such as calibration, diagnostic, and control require low latency and high reliability to meet the strict in-vehicle control system specifications in real time. The chassis and safety data transmission also require high reliability so that the ADAS works efficiently. If the ADAS data transmission is unreliable, then it might result in severe damage to the vehicle and the driver. The ADASs are based on complex algorithm, and the algorithm can be built easily without modifying the sensors. They have to be capable of performing over-the-air (OTA) updates. Nevertheless, the infotainment subsystems such as display, multimedia, diagnostics, and connectivity do not require very high reliability compared to the powertrain and chassis subsystems. However, it requires a medium reliability to operate efficiently.

4.6.2 In-Vehicle Onboard Ports, Threats, and Countermeasures

There are a number of ports inside modern vehicles that, when linked to external devices, allow users to access services and infotainment information. This section discusses three vehicular ports and their security threats. The 3 vehicle ports are onboard diagnostics (OBD)-II ports, the USB ports, and electric vehicle charging ports. Such ports synchronize user's smartphones and recharge their electric cars. But, if hackers obtain access to such ports, they might get access to the in-vehicle system, execute eavesdropping attacks, and perhaps even install malware and ransomware. As previously stated, the different types of attack vectors through which attackers may obtain access to the internal networks of a vehicle are the physical access and the various wireless interfaces found in the vehicles. Such interfaces allow external inputs from which core ECUs can be exploited remotely by software bugs, remote control of a car over the Internet, among others.

4.6.2.1. OBD-II ports: The OBD-II port is an onboard device that monitors the vehicle's emissions, mileage, speed, and other details. It tracks various modules or subsystems for emissions and engine components and can illuminate the malfunction indicator lamp (MIL) in case of malfunction. The OBD adapter provides direct connectivity to all CAN buses inside the vehicle compartment through a physical connection. The OBD-II features a 16-pin port situated under the side dash of the driver. A technician or someone uses special search tools to decipher the error message. OBD-II ports are weak part of vehicle safety, as linking to the OBD-II port allows diagnostic data to be collected and entry to the in-vehicle network and malicious program deployment. There are two types of threats faced by OBD-II port, and they are

a. In-vehicle network access attack, where the attacker can install a malicious equipment in the vehicles' internal network through the OBD-II port to obtain physical access.
b. Dongle exploitation attack: In this type of attack, dongles are inserted into the OBD-II ports and it can be handled remotely and can be easily decrypted. Such dongles, which are attached to the OBD-II ports, are vulnerable and could be compromised when a brute-force attack is carried out by the hackers and enabled them to send harmful messages over the CAN that led to vehicle's engine failure.

4.6.2.2. USB Port: The USB ports gain popularity and are installed in all the modern vehicles because they can communicate with computers, GPS systems, and USB accessories to the vehicle. The use of USB ports in the vehicle poses additional security threats. The hackers could use the USB ports to reprogram the controller processor, install malicious codes, tamper network cards, and finally target the operating system.

4.6.2.3. Electric Vehicle (EV) Charging Port: The EVs need to be charged frequently as compared to the conventional vehicles that use fuels. EVs would need to be charged more often than petrol vehicles need to be fueled at the gas station. EVs would be susceptible to attacks via the charging infrastructure when charging. The charging system could potentially be used for carrying out attacks on a smart grid. The malicious software launched by the attackers might compromise the EV and then compromise the charging station itself. The attacker can compromise the location privacy by detecting the charging station entry and exit information.

4.6.2.4. Countermeasures to Port Threats: In IVN, the OBD-II threats can be overcome by tracking the frame injection coming from the OBD-II port as well as encrypting and signing the message for firmware updates. Due to the large application of USB devices, it poses security vulnerabilities so a standard USB security should be developed to protect against the cyber-attacks. The threats to USB port can be prevented in two ways; first, the USB that connects to the Internet should require a security certificate, which may then permit it to link to the vehicle, and secondly, preventing the access of malware or viruses to enter the sensitive security area via the USB port. To prevent the EV charging port attack, authentication schemes, cryptographic signatures, and secure firmware updates should be used to prevent the various attacks during charging. In addition, Open Charging Point Protocol (OCPP)-based secure charging system in smart grid can be used to secure the EV charging systems.

4.7 Cyber Security in In-Vehicle Network (IVN)

4.7.1 In-Vehicle Network (IVN) Security Threats

In the past, the vehicles were designed for comfort and safety, not for security, so attackers easily take control of the vehicles using physical attacks. Due to the development of the automobile industry, cybersecurity is a non-trivial problem for security experts. Currently, there are several companies, which are working to develop safety and security protocols to safeguard the IVN. The research into IVN bus security has grown recently, mainly due to several weaknesses in security of existing IVN. The hackers were able to monitor as well as control various vehicle functions using the IVN bus network and reverse engineering the ECU's code, such as locking the engine and disabling the brakes. CAN-based cybersecurity offerings focus on protecting the bus from compromised ECUs. While these will still be the core threat to consider with Ethernet networks, the physical gateway network architecture and virtual segmentation will raise new issues. It is possible to exploit the IVN bus system with the help of wireless networks (e.g., Bluetooth, cellular, and Wi-Fi networks), without any kind of physical access to the vehicle. Researchers like Miller and Valaske showed the successful attacks utilizing the vulnerabilities of CAN bus network on Toyota Prius and Ford vehicle. They hacked and manipulated the braking system, PS, speedometer, steering, and other ECUs. In addition, they remotely disabled the braking system of Jeep while driving [17]. Keen Technology Lab posted an early hack on Tesla, in which hackers control the vehicle by taking advantage of a flaw from the infotainment system's browser, causing the company to release a software update over the air. Similarly, automotive Ethernet (AE) security strategies must involve more than just the detection/override and drop/redirection of malicious signals. Compared with CAN techniques, handling rogue messages or ECUs properly and effectively means considering the specifics of the network, its architecture, protocols, and applications. Hereafter, we will discuss the in-vehicle security threats on the IVN protocols one by one.

- 6.7.1.1. CAN Security Threats: CAN bus system is vulnerable to different types of attacks such as injection, masquerading, denial of service (DoS), eavesdropping, replay, and bus-off attacks [18]. Masquerading attacks occur in CAN because the CAN frames are not encrypted. Thus, the attackers get knowledge of the CAN frames and find entry points of the network. In addition, the CAN fails to support message authentication. The attackers eavesdrop the broadcasted vehicular CAN messages and then gain access to the in-vehicle networks. During injection attack, the attackers insert bogus signals into a vehicle bus system. The attackers obtain access into the in-vehicle system via OBD-II ports and compromise the ECUs and infotainment or telematics systems. During a replay attack, the attackers constantly resend legitimate frames to hinder real-time operation of the vehicle. In bus-off attacks, the attackers send bits constantly in the identification field as well as in other fields, forcing the Transmit Error Counter (TEC) which then turns off the ECU once the TEC value becomes greater than 255. In case of DoS attacks, the attackers deliver high-priority CAN packets constantly that obstruct valid low-priority packets and hamper the normal operation of in-vehicle communication. As a result, the attackers finally take control of the vehicle. For countermeasures against attacks such as masquerading, eavesdropping, injection, and replay attacks, the messages exchanged between ECUs could be encrypted and authenticated or by transmitting a fraction of MAC in each frame such that the tampering recognition can be carried out for all separate frames [18].

6.7.1.2. FlexRay Security Threats: In FlexRay, the most common types of threats are eavesdropping attacks and static segment attacks. In case of eavesdropping attacks, the attackers gain access to the FlexRay messages and obtain all the information. It causes data leakages, affects data confidentiality, and concerns security primitives. The static segment attacks include masquerading, injection, and replay attacks, and the attackers attack the static segment of the FlexRay communication cycle. The countermeasure for both the static segment attacks and the eavesdropping attacks is by implementing message authentication within the static segment via hardware coprocessor, where feasible. Security-Aware FlexRay Scheduling Engine (SAFE), which is a FlexRay scheduling system, implements the authentication protocol know as Timed Efficient Stream Loss-tolerant Authentication (TESLA) for security [19].

6.7.1.3. LIN Security Threats: LIN is commonly used along with CANs and is therefore prone to unwanted access through the CAN bus. The master–slave model of LIN implies that CAN-specific vulnerabilities do not necessarily extend to LIN. Some of the attacks to LIN are response collision, message spoofing, and header collision attacks. The vulnerabilities in LIN master–slave model cause message spoofing attacks where the attacker sends unauthorized messages with false information to shut down the LIN interrupting vehicular communications. During a collision response attack, an attacker concurrently transmits an illicit message with false header along with a valid message to exploit LIN's error handling protocol. As a result, the legitimate slave node of the LIN stops the message transmission, while other nodes will consider the illegitimate message to originate from authentic nodes. In header collision attack, the attacker transmits incorrect header to conflict with a valid header sent by the master node. The valid header states that a certain slave node must release the response; however, the attacker's collision implies that the source node has changed. This attack causes unwanted functions such as keep automated sliding doors open and sometimes lock the steering wheel while the vehicle is driving. The countermeasure for these types of attacks is to send unusual signals by the slave node so that it will overwrite the attacker's fake messages, when the bus value mismatches its response.

4.7 Cyber Security in In-Vehicle Network (IVN)

6.7.1.4. AE Security Threats: The threats to AE security are traffic integrity attacks, network access attacks, DoS attacks, and traffic confidentiality attacks. In case of network access attacks, the attacker connects to the Ethernet network by connecting to the unsecured port of the switch, gains access to the network, takes control of other nodes or switches, or remotely accesses to the Ethernet network. In traffic confidentiality attacks, when attackers have accessibility to the network, they may perform security attacks on networks, enabling them to eavesdrop on network activity. Traffic integrity attacks may be similar to man-in-the-middle (MITM) attacks that divert network traffic to the intruder's node to exploit the information. Examples of traffic integrity attacks are session hijacking attacks and replay attacks. There are two types of DoS attack in Ethernet, i.e., layer 1 attack that physically destroys links or hardware, leaving the Ethernet infrastructure entirely unusable, and layer 2 attack that is resource depletion attacks, which waste energy by submitting frames to be analyzed constantly, or protocol-based DoS attacks. The countermeasures to these types of attacks are authentication approach and frame replication method that eliminate traffic confidentiality and traffic integrity attack. The network access attack can be countermeasured by virtual local area network segmentation.

6.7.1.5. MOST Security Threats: There are two common types of attacks in MOST, and they are synchronization disruption attacks and jamming attacks. In synchronization disruption attacks, the attacker sends the fake timing frames to tamper the synchronization of MOST. In jamming attacks, the jammer repeatedly delivers misleading messages to interrupt legitimate low-priority specified-length messages or constantly requesting data channels on MOST transmission through control channels. The countermeasures on MOST security can be achieved by authenticating the source nodes, encrypting the exchanged messages, and implementing firewalls and gateway.

4.7.2 Cybersecurity Protection Layers

In this section, we argue that the cybersecurity issues must be resolved at the earliest possible time to detect vulnerabilities rather than later. Otherwise, it might lead to severe problem and jeopardize vehicles on the road. This is the era of connected and autonomous vehicles with complex and sophisticated vehicle architecture. The connected and autonomous vehicles consist of a variety of sensors, computer systems, electronic devices, etc. They need robust cybersecurity to guarantee that these systems work properly and are designed to reduce security risks.

There have been few works done to secure the IVN bus and gateway systems. A security layer has been proposed at the infotainment device by introducing secure gateway (SG) concept, which provides protected access to the internal vehicle network from installed applications in the infotainment system. Three secure layers have been used to protect the CAN bus system; they are network layer, messaging layer, and service layer [20]. Similarly, Intel discussed 15 key vulnerable features prone to attack in their white paper [21]. According to Intel, the security of dynamic and complex systems of smart vehicles requires coordinated, systemic solutions. In addition, successful security cannot be accomplished by reacting to the threats or attacks of individual parts, as opposed to conventional computer systems. This makes it more difficult to secure vehicle systems as they can attack the vehicles in cyber as well as in physical world. From this perspective, the safety and security of vehicle systems must adhere to the Framework ver. 1 for the cyber-physical system described by NIST.

In case of IVN cybersecurity, a multilayered strategy is required to maintain a robust cybersecurity environment that leverages existing cybersecurity frameworks and enables organization to follow guidelines that improve the vehicle's overall security. In addition, it is essential to adopt cybersecurity by focusing on the entry points of a vehicle, both wireless and wired, which could potentially be vulnerable to cyberattacks.

We present a multilayered strategy for the cybersecurity of vehicles to minimize vulnerability in cyberattacks and reduce the possible effects of a successful intrusion as shown in Fig. 4.6. The autonomous and intelligent vehicle and the existence of hackers are now part of life—security must therefore be an integral part of the autonomous and intelligent vehicle design, as security is as fragile as the weakest contact. Thus, the protection for the electrical system and infrastructure for vehicles is important to ensure safety for vehicle passengers. A detailed and systematic approach to design multilayered vehicle cybersecurity protection for in-vehicle system is shown in Fig. 4.7.

The five-layer security system offers a holistic solution for protecting the whole infrastructure of in-vehicle system.

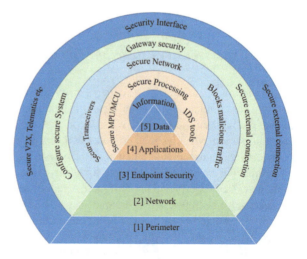

Fig. 4.6 Cybersecurity protection layers for autonomous vehicles

4.7 Cyber Security in In-Vehicle Network (IVN)

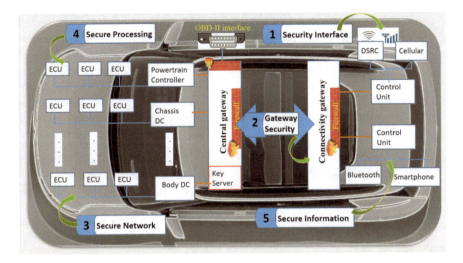

Fig. 4.7 In-vehicle cybersecurity protection layers

1. Security Interface: The security interface layer is the first layer that secures the external connection of the autonomous vehicles by limiting physical access to the vehicle control system. This layer protects the communication channels against the data manipulation and data theft by authenticating and validating the messages that are exchanged to protect the message integrity. It also secures the V2X communication, telematics, and other infotainment systems. The attackers can hack the TCU by sending spoofed CAN messages to access control of critical safety units such as steering and breaks. This layer adds security to the TCU and OBD by attaching a secure element to provide maximum protection against attacker. Secure elements are dedicated protection microcontrollers with high-performance cryptographic accelerators, and these elements demonstrated improved resistance to physical and electrical attacks.
2. Gateway Security: The gateway firewall is the second layer that helps to shield the safety-critical functions. The gateway ensures the system is configured securely. The gateway is a context-aware routing function that decides which messages are currently valid and therefore be transferred through the gateway to the destination via a variety of increasingly complex controls. The central gateway ECU isolates the TCU/OBD from the network and separates the vehicle network into functional domains [22]. In case of Jeep hack, the Jeep did not have any gateways and domain isolation associated with it, so that hackers disabled the brakes remotely [22]. A central gateway performs several functions connecting data to signals from the different nodes around the car for the conversion of a variety of automotive protocols. It provides domain separation between the infotainment systems and critical safety systems. The secure gateway is built on standard encryption IP protocols, and the network adapters are used to interact with CAN bus networks.

3. Secure Network: The secure network layer is the third layer that delivers secure connection between the ECUs. The secure network layer secures the transceivers by implementing security features and keeps the ECU communication private and reliable. The gateway security layer reduces the attack surfaces significantly by separating the vehicular networks into functional domains. However, the subdomains are still susceptible to attack surfaces. Thus, this layer guards the subdomains by using a secure network composed of four security measures and they are as follows:

   a. ECU-level validation: The validity of ECUs can be checked periodically in the network such as on engine start as well as occasionally afterward.
   b. Encryption: Data loss and identity theft can be prevented by encrypting messages inside the vehicle, which are shared between various ECUs.
   c. Message authentication system: Each message is augmented with a cryptographic certificate to ensure an authenticated sender, as well as preserve the integrity of the message.
   d. Detection of intrusion: Guidelines and recognition of unusual pattern for identifying irregularities in network traffic and blocking suspicious packets before they can possibly reach the microcontroller.

   These security measures can be enabled by microcontroller integrated security subsystems including cryptographic accelerators.

4. Secure processing: The secure processing layer is the fourth layer that implements all connected and autonomous vehicles' features. We need to guarantee authentic and reliable software running on the central processing unit (CPU). Modern microcontrollers include safe booting and real-time integrity testing schemes to ensure that the encryption is legitimate, trustworthy, and unchanged. Mechanisms for regulated lockout of the MCU and ECU are used to lock out debugging and software download features that would be invaluable for hackers. The autonomous vehicle consists of millions of codes and several microcontrollers that makes the vehicle software complex, and as a result there exist some vulnerabilities. When any security vulnerability or bug in the software is detected, the vehicle software should be securely, quickly, and seamlessly updated without the requirement to visit the vehicle maintenance store or garage. Thus, a secure online software update system such as over-the-air (OTA) software updates is required for every ECU in the autonomous and connected vehicles.

5. Secure Information: The secure information layer is the fifth layer in the cybersecurity protection layer for the autonomous vehicles. The ECUs generate, process, share, and store large volumes of potentially sensitive information continuously making them potential targets for hackers. The integrity, confidentiality, and reliability of user data must also be maintained. As a general basis for ensuring adequate data protection, the user personal information can be anonymized by using pseudo-anonymization techniques. The automotive microcontrollers and processors include integrated automotive security modules like secured hardware extension (SHE) and hardware security module (HSM) that protect information

from being manipulated, and support stable software updates and data protection. Safe access to vehicles is the conventional side of vehicle security, including solutions for vehicle access.

The reference cybersecurity layer might vary depending on advanced cyber defense technologies. While there are several technical solutions to protect against known cyber-attacks, there is a less well-defined guideline to combat unknown threats.

4.7.3 Cybersecurity for ECU

In this section, we discuss the cyberattacks in ECUs including attack vectors, attack surfaces, attack types, and its countermeasures. The ECUs are vulnerable to attacks that compromise the vehicles due to the lack of strong security in ECUs. The attackers or hackers gain control of the vehicles or extort the vehicle owner by installing ransomware that disables the functioning of the vehicles. Thus, to secure the ECUs against the attackers, a strong security measure is required to prevent unauthorized access, theft, information disclosure, impact operation, etc.

The attack vector in ECU is a one-way path by which the attackers gain unauthorized access to the ECU and its features. The multiple attack vectors in the ECU systems together form the attack surface, where the attackers attempt to attack the system through the ECU interfaces. The ECU has two interfaces, i.e., user channels and restricted channels as shown in Fig. 4.8. The user channels are used for normal operation such as communication interfaces for the IVNs through wired communications or inter-vehicle networks through wireless communication, while the restricted channel by its name means to use for privileged operations such as internal test, maintenance, and debug. The hackers can use these weak interfaces as the attack vector using different attack types such as eavesdrop, replay, and fake message attack. It is necessary to protect the above-mentioned channels and interfaces from being part of the hacker's attack surface.

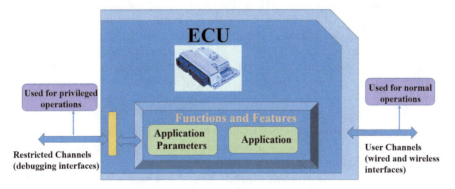

Fig. 4.8 ECU channels and interfaces

Table 4.4 Types of attack classes at various levels in an ECU

Entry point	Attack class	Attack types	Countermeasures
Remote	On communication	Eavesdrop, replay, fake message insertion, flooding	Encryption, authentication, rate limitation
	On exposed functions	SW bugs, known HW bugs, resource exhaustion, design flaws	Code analysis, host IDS, watchdogs, firewalls
Local	Noninvasive	Timing analysis, EMA analysis, photoemission analysis	Execution jitter, constant time execution, constant power execution
	Semi-invasive	Local light attacks, spike, glitch, signal injection, alpha particle penetration	Light sensor, sensors on supply or signals, information redundancy
	Invasive	Micro-probing, PCB modification, AFM, delaying	Camouflage, shield, masking, tamper detect pin

The degree at which an attacker has access to the ECU is one of the key factors that determines the extent of such attacks. The attackers can have access to the ECU based on various entry point levels, such as remote or local access levels. In case of remote access level, the attacker can attack the ECU system through the vehicle's external wireless interfaces rather than the IC chip or PCB level, while in case of local access level, the attackers can get local access to the ECU through its user and restricted interfaces. In addition, it can get local access to the PCB board/IC chip through the internal wiring and test points. There are five types of attack classes at various levels in an ECU as shown in Table 4.4. The countermeasures to these attacks are also included in the table, which is not an extensive list of countermeasures, but relatively suggestion of measures that can be applied per types of attacks. The five types of attacks are described below.

On Communication: The communication attacks are caused by attackers that actively listen into the private communication between the system modules, or exhausting the BW in a channel. Some of the communication attacks are eavesdropping, flooding, man-in-the-middle (MIMT), replay, fake messages, etc. The fact that certain countermeasures were or were not implemented somewhere else in the VEE system may have an effect on the particular requirements per ECU. The countermeasures to these attacks are encryption, authentication, and rate limitation.

On Exposed Functions: The attacks on exposed function are another type of remote attacks that occurs through the interfaces on the weakly implemented exposed function of the target parts. These attacks occur due to the design flaws, software (SW) and hardware (HW) bugs, and resource exhaustion due to buffer overflows due to the weak implementation. These attacks can be countermeasured by code analysis, host IDS, watchdog mechanisms, firewalls, etc.

Noninvasive: The noninvasive attack is a local access-level attack that occurs when the ECU and its components leak the important information through side channel. It is also known as side channel attack (SCA) that can be used to observe the internal behavior of the system. Some of the noninvasive attacks are timing analysis, electromagnetic analysis (EMA), photoemission analysis, etc. The countermeasure to prevent the noninvasive attacks is execution jitter, constant time execution, and constant power execution.

Semi-Invasive: In case of semi-invasive local access attacks, the attacker uses the local light attacks (using flashlight or laser beams) that might introduce extra charge inside an ICU to momentarily modify the state of the ECU or its components. The attackers might introduce the spike or glitch in the source voltage or signal that cause the unintentional behavior or imbalances. To countermeasure, the semi-invasive attacks, light sensor, sensors on supply or signals, information redundancy can be used that minimizes the effect of the attack severity.

Invasive: In invasive attacks, the attackers permanently change the system design by changing the PCB interconnection or changing the circuit of the IC using PCB modification schemes like focused ion beam. Such types of attacks are micro-probing, PCB modification, AFM, delaying. In order to countermeasure the invasive attack, methods such as tamper detect pin, camouflage, shield, masking can be implemented.

4.8 Summary

In this chapter, we outlined the in-vehicle communication system including various types of sensing technologies used in automotive vehicles. In order to provide safety, security, and comfort to the customers, various electrical and electronics systems are installed in the vehicles. Presently, the number of such electronic devices for vehicle safety and infotainment has increased steadily and continuously. Even though the in-vehicle communication system enables many new services, it also exposes the vehicle and its software to potential attacks and threats. Thus, chapter presents the in-vehicle communication system and cybersecurity issues. The following main topics are highlighted: in-vehicle electrical and electronic system, and in-vehicle communication systems which include various types of sensing technologies such as sensing, vision, and positioning technologies. We discuss the different types of in-vehicle networking (IVN) systems and their security threats. Then, we present the IVN architecture and challenges of IVN architectures. Finally, we discuss the cybersecurity in IVN and its security protection layers in detail.

References

1. D.P.F. Möller, R.E. Haas, *Guide to Automotive Connectivity and Cybersecurity* (2019)
2. H. Abraham, H. McAnulty, B. Mehler, B. Reimer, case study of today's automotive dealerships: Introduction and delivery of advanced driver assistance systems. Transp. Res. Rec. **2660**(1), 7–14 (2017)

3. N. Navet, F. Simonot-Lion, *In-Vehicle Communication Networks—A Historical Perspective and Review* (Luxembourg, 2013)
4. W. Zeng, M.A.S. Khalid, S. Chowdhury, In-vehicle networks outlook: achievements and challenges. IEEE Commun. Surv. Tutorials **18**(3), 1552–1571 (2016)
5. B. Groza, S. Murvay, A. van Herrewege, I. Verbauwhede, *LiBrA-CAN: A Lightweight Broadcast Authentication Protocol for Controller Area Networks BT—Cryptology and Network Security* (2012), pp. 185–200
6. BOSCH, *CAN With Flexible Data-Rate Specification* (2012)
7. Vector Informatik GmbH, *Testing ECUs and Networks with CANoe*. [Online]. Available: https://www.vector.com/int/en/products/products-a-z/software/canoe/
8. C.P. Quigley, R. McMurran, R.P. Jones, P.T. Faithfull, An investigation into cost modelling for design of distributed automotive electrical architectures, in *2007 3rd Institution of Engineering and Technology Conference on Automotive Electronics* (2007), pp. 1–9
9. G. Han, H. Zeng, Y. Li, W. Dou, SAFE: Security-aware FlexRay scheduling engine, in *2014 Design, Automation and Test in Europe Conference & Exhibition (DATE)*(2014), pp. 1–4
10. C. Patsakis, K. Dellios, M. Bouroche, Towards a distributed secure in-vehicle communication architecture for modern vehicles. Comput. Secur. **40**, 60–74 (2014)
11. D. Porter, *100BASE-T1 Ethernet: the evolution of automotive networking* (Dallas, Texas, 2018)
12. W. Jun Hyung, Automotive digital measurement solution. Keysight. [Online]. Available: https://www.keysight.com/main/eventDetail.jspx?ckey=2957114&id=2957114&nid=-11143.0.08&lc=eng&cc=US. [Accessed: 02-Apr-2020]
13. V.M. Navale, K. Williams, A. Lagospiris, M. Schaffert, M.-A. Schweiker, (R)evolution of E/E architectures. SAE Int. J. Passeng. Cars Electron. Electr. Syst. **8** (2), 282–288 (2015)
14. S. Dhaneshwar, *Automotive Cyber Physical Systems*. [Online]. Available: https://www.linkedin.com/pulse/automotive-cyber-physical-systems-shashank-dhaneshwar/. [Accessed: 03-Apr-2020] (2017)
15. C. Ebert, E. Metzker, Functional safety and cyber-security—experiences and trends, in *Functional Safety Symposium* (2018), pp. 1–25
16. J. Huang, M. Zhao, Y. Zhou, C. Xing, In-vehicle networking: Protocols, challenges, and solutions. IEEE Netw. **33**(1), 92–98 (2019)
17. C. Miller, C. Valasek, remote exploitation of an unaltered passenger vehicle. Black Hat USA **2015**, 1–91 (2015)
18. Z. El-Rewini, K. Sadatsharan, D.F. Selvaraj, S.J. Plathottam, P. Ranganathan, Cybersecurity challenges in vehicular communications. Veh. Commun. **23**, 100214 (2020)
19. A. Perrig, R. Canetti, J.D. Tygar, D. Song, The TESLA broadcast authentication protocol. RSA CryptoBytes Tech. Newsl. **5**(2), 2–13 (2002)
20. J. Berg, J. Pommer, C. Jin, F. Malmin, J. Kristensson, Secure gateway-a concept for an in-vehicle IP network bridging the infotainment and the safety critical domains, in *13th Embedded Security in Cars (ESCAR'15)* (2015), pp. 1–12
21. D.A. Brown et al. *Automotive Security Best Practices* (2015)
22. A. Birnie, T. van Roermund, *A Multi-Layer Vehicle Security Framework* (2016)

Chapter 5
AUTOSAR Embedded Security in Vehicles

5.1 Overview

We live in a world that is getting more interconnected by each day, and we are witnessing a global change where all the devices in our surroundings are becoming "smart" and connected to the Internet. The automotive industry is also a part of this change. Today's vehicles have more than 150 small computers, embedded control units (ECUs), and multiple connection points to the Internet which makes them vulnerable to various online threats. Recent attacks on connected vehicles have all been results of security vulnerabilities that could have been avoided if appropriate risk assessment methods were in place during software development. In this chapter, we demonstrate how the threat modeling process, common for the computer industry, can be adapted and applied in the automotive industry. The overall contribution is achieved by providing two threat modeling methods that are specifically adapted for the concept of the connected car and can further be used by automotive experts. The methods were chosen after an extensive literature survey and with the support of domain experts from the vehicle industry. The two methods were then successfully applied to the connected car and the underlying software architecture based on the AUTOSAR standard. We have empirically validated our results with domain experts as well as tested the found vulnerabilities in a simulated vehicle environment.

Adi Karahasanovic and Pierre Kleberger
Combitech AB, Lindholmspiren 3A SE-417 56 Gothenburg Sweden
E-mail: adi.karahasanovic, pierre.kleberger@combitech.se.

Magnus Almgren
Department of Computer Science and Engineering,
Chalmers University of Technology, SE–412 96 Gothenburg, Sweden
Email: magnus.almgren@chalmers.se.

5.2 Introduction

The world as we know it is changing and many of the devices we use daily are becoming "smart." This buzzword is appearing in everything from smart grids to smart homes with the smart appliances therein. The main aspect of these devices is their connection to the Internet, and because of it, previously local vulnerabilities are now widely exposed. The same goes for many new vehicles. A high-end car may now have more than 100 million lines of code [1], as well as multiple connections to external networks including the Internet. All this code has to be properly developed and tested in order to ensure the safety and security of the vehicle. We will refer to these types of cars as connected cars.

The connected cars are equipped with a number of new technologies and features that are not possible without an Internet connection. For example, drivers now have the possibility to receive service information and traffic reports through the vehicle's dedicated cellular connection. They also have the possibility to connect their smartphone to the vehicle (Bluetooth, Wi-Fi) and use its Internet connection to enable some of the new features of the vehicle's entertainment center. Through this center, they can browse the web, access social networks, stream online content, etc. Many other services exist depending on the vehicle type and manufacturer. According to the Business Insider report, there will be 380 million connected cars on the road by the year 2021 [2].

Even though the connection to the outside world enables many new services, it also exposes the car and its software to a potential remote attack. There have already been a number of successful cyberattacks on connected vehicles such as the attacks on Jeep Cherokee [3], Tesla S model [4], Nissan electric car [5], and Chevrolet Corvette [6]. As the production of these vehicles increases, so will the importance of securing them. We argue that it is important to address these security concerns as early as possible. Using threat modeling methods already at the design phase assists in early detection of security flaws instead of later detection, which may lead to a recall of many cars already on the road. Another benefit of our study is to provide a common framework that any car manufacturer can use.

In this chapter, we investigate two threat modeling methods widely used in the computer industry and evaluate their suitability for the connected car. We also propose adaptations to these threat modeling methods to make them more applicable to the underlying software architecture used in today's vehicles (AUTOSAR). The first method, TARA, represents an attacker-centric approach while the second method, STRIDE, investigates the software architecture of the system, and belongs to the software-centric approach.

The outline of this chapter is as follows. In Sect. 5.2.1, we summarize the two main threat methods that are used, as well as the automotive standard AUTOSAR. The adaptation of the threat methods to the automotive domain is then described in Sect. 5.3. The process of applying the adapted version of these two methods is described in Sect. 5.4. Even though security specialists have vetted the modified frameworks, we add an empirical verification of the results from the threat analysis

5.2 Introduction

by testing the found vulnerabilities in a simulated vehicle environment. The results are described in Sect. 5.5. We summarize related work in regards to threat modeling in Sect. 5.5.3. Section 5.5.4 presents a discussion of our results, followed by our conclusions in Sect. 5.5.

5.2.1 Background

After completing a larger survey of relevant literature, we chose two particular threat modeling frameworks popular in the computer industry (TARA and STRIDE) that seemed to be most suitable for the automotive industry. In this section, we give a high-level overview of these frameworks before we discuss our suggested changes in Sect. 5.4. We also describe the AUTOSAR standard, as it is a central concept in the chapter.

5.2.1.1 TARA

The threat agent risk assessment (TARA) method was developed by security experts from Intel Security [7] and is based on three groups of collected data, denoted as libraries:

- **Threat Agent Library (TAL)** — lists all relevant threat agents and their corresponding attributes.
- **Methods and Objectives Library (MOL)** — lists methods that each threat agent might employ along with a corresponding impact level.
- **Common Exposure Library (CEL)** — lists areas of the greatest exposure and vulnerability.

These libraries are populated internally inside a company by their security expert team. They are based on incident reports, breach reports, security measures, and other confidential information that is required to create the libraries. By using the information from the libraries, security experts can determine which threat agent attributes are needed in order for a threat agent to pose a threat to the company and its assets. The information is also used to list the methods that are most frequently used to attack these assets, along with a list of the most exposed areas. Each of the exposed areas is described with the level of exposure, current security control that protects this area, and the recommended security control for this area. The difference between the current and the recommended security control shows which area needs a new or improved security measure. By combining the information from all three libraries, security experts can determine which areas are most likely to be attacked with what method and by which threat agent. To conclude, these libraries provide the relevant information that is required in order to align the security strategy and thus target the most important exposures. Once derived, these libraries can be updated and used in future instances o f he TARA method.

5.2.1.2 STRIDE

The STRIDE method was originally developed by Microsoft [8]. The method allows threat identification in the design phase of any software or hardware and as such gives insight into potential attack scenarios. There are two variants of the STRIDE method: per-interaction and per-element. In order to apply the method, security experts first need to create data flow diagrams (DFD) of the system that needs to be analyzed. The DFDs present the communication patterns between the components under investigation. Afterward, the method examines these diagrams in order to detect possible threats to the system. The threats are divided into six different categories: spoofing, tampering, repudiation, information disclosure, denial of service, elevation of privileges. The inspection of DFD diagrams can be done manually (brainstorming sessions) or by using the Microsoft Threat modeling tool which uses the STRIDE per-interaction variant. The method described in this chapter (further described in Sect. 5.3.2) uses the MS Threat modeling tool (version 2016). As a result, the tool generates a complete list of all found threats based on the input DFD diagram.

5.2.1.3 AUTOSAR

AUTomotive Open System Architecture (AUTOSAR) was founded in 2003, with the goal to develop an architecture, independent of the underlying ECU hardware that the automotive industry can use to reduce the increasing complexity of software in modern vehicles [9]. It is the de facto standard for the automotive software today, and 80% of global production is based on this standard.

AUTOSAR makes an abstract layer of the underlying hardware, so that the applications written on top of AUTOSAR are independent from the actual supplier of the ECU hardware. The AUTOSAR standard defines security mechanisms that can be used by the software modules implemented into the vehicle system. It further specifies interfaces and procedures to provide Secure On-Board Communication, and the exact implementation is left for the OEMs to decide on. OEMs choose the cryptographic algorithms and encryption techniques which they want to implement and use in the vehicle system [10]. The three main security mechanisms in the AUTOSAR are Crypto Service Manager (CSM), Crypto Abstraction Library (CAL), and Secure On-Board Communication (SecOC).

A. Software Architecture

The AUTOSAR standard documentation guides companies and the automotive industry in designing and implementing software in their vehicles. By adopting the AUTOSAR standard, companies can develop software solutions that are independent of the hardware they are running on, and this software can run on any ECU in the vehicle. This is the reason why the AUTOSAR platform is also called a hardware-independent architecture. Figure 5.1 shows the difference in the vehicle software architecture when the AUTOSAR standard is adopted.

5.2 Introduction

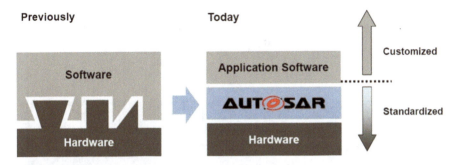

Fig. 5.1 AUTOSAR hardware-independent architecture [11]

Besides defining the architecture and the interfaces, this standard also defines a design flow that specifies how software should be mapped to the ECUs during the development cycle. This process ensures that all the companies use the same approach for implementing this standard into their vehicles. The high configurability of the AUTOSAR implementation allows it to be adjusted to the specific needs of each company and guides the application software that each company is developing for their specific vehicles. This is one of the main goals of this standard, to use the knowledge and experience of every member company in order to improve and further develop the standard on one hand, while on the other hand allow each company to develop its own software applications and configure the AUTOSAR implementation to their own needs. This allows companies to "cooperate on standards, compete on implementation" which is the official AUTOSAR motto [12, 13].

AUTOSAR provides detailed specification for:

- Software architecture
- Software development methodology
- Standardized application interfaces (APIs) [11].

Figure 5.2 shows the three-layered architecture of the AUTOSAR standard: application layer, run-time environment (RTE) layer, and the basic software (BSW) layer that consists of four sublayers. Each of the sublayers offers different services as shown in Fig. 5.3. The highest layer is the application layer, which contains the software components (SWCs). AUTOSAR application (e.g., ABS or the seat heating control) consists of several SWCs, which provide the core functions that are used by the AUTOSAR application. The AUTOSAR SWC is an atomic piece of software that cannot be divided and is located on one ECU. The SWCs used in this thesis are sensor and actuator SWCs.

The sensor SWC reads the state of the sensor and provides that data to other components or the AUTOSAR application, while the actuator SWC sets the state of an actuator located on that ECU. SWCs are ECU dependent while the AUTOSAR application is not dependent on one ECU. The interaction between the AUTOSAR application and the AUTOSAR SWCs is achieved through the RTE layer [14].

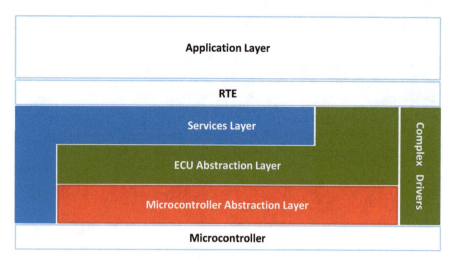

Fig. 5.2 Layered AUTOSAR architecture [11]

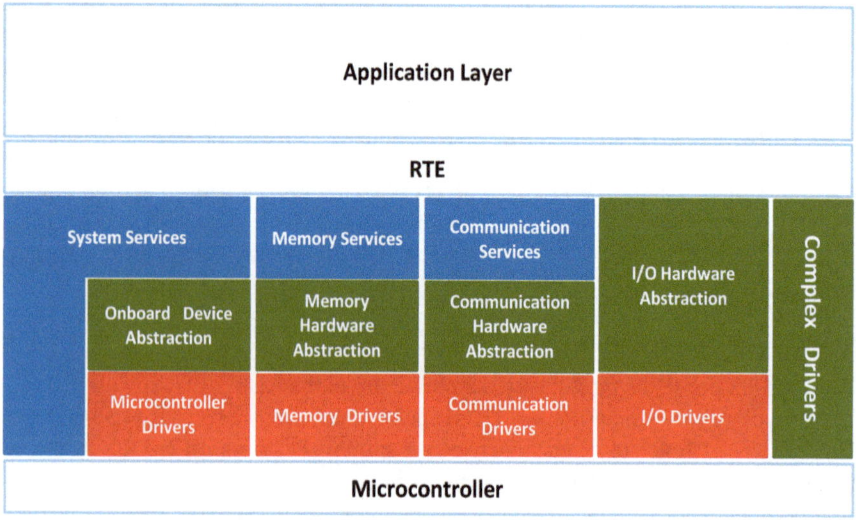

Fig. 5.3 Each sublayer of the BSW layer offers different services [11]

The run-time environment (RTE) layer provides communication ability between the SWCs and the basic software layer. Because of this layer, SWCs can be used on different ECUs, independent of the ECU vendor. This layer is ECU specific because the specification of the ECU determines the correct access to the right communication channel [12, 15].

5.2 Introduction

Basic software layer (Fig. 5.2) consists of [12, 15]:

- **Services layer**—This is the highest layer and provides basic services for applications, RTE and BSW modules. It offers operating system functions, i.e., communication services, memory services, diagnostic services, ECU state management, and logical and temporal program flow monitoring. The layer also contains the main security mechanism of the AUTOSAR standard: CSM cryptographic module and the SecOC security module, which are explained in more detail in subsection B of AUTOSAR.
- **ECU abstraction layer**—This layer responds to the functions of the applications and interacts with the drivers of the microcontroller abstraction layer. It provides an application programming interface to devices regardless of their location (internal/external microcontroller) including external devices. This layer makes the higher layers independent of the ECU layout.
- **Microcontroller abstraction layer**—This layer contains drivers for direct access to the microcontroller and internal parameters. It makes higher layers independent of the microcontroller.
- **Complex drivers layer**—The layer provides the ability to integrate special-purpose functions such as drivers for devices that are not specified with the AUTOSAR standard. This layer accesses the microcontroller directly.

Figure 5.4 illustrates a more detailed view of the AUTOSAR software architecture including the paths of internal communication between different layers and internal services. One new concept shown in Fig. 5.4 is the AUTOSAR Virtual Functional Bus (VFB), which defines a communication framework between SWCs. VFB abstracts the communication away from the hardware in such a way that it does not matter if two communicating SWCs are on the same ECU or on different ECUs [16].

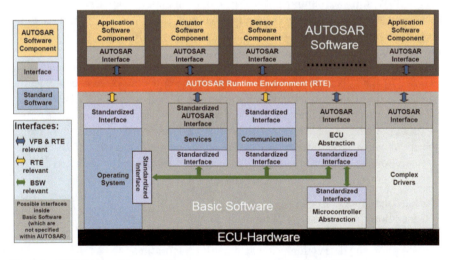

Fig. 5.4 AUTOSAR software architecture—components and interfaces [11]

B. Security features

The AUTOSAR standard defines important security mechanisms that are used by the SWCs and other software modules implemented into the vehicle system. It further specifies interfaces and procedures for ensuring Secure On-Board Communication while the rest of the security mechanisms and implementations are left for the OEMs to decide on. OEMs choose the cryptographic algorithms and encryption techniques that they want to implement and use in the vehicle system [17]. The three main security mechanisms in the AUTOSAR are:

- CSM - Crypto Service Manager
- CAL - Crypto Abstraction Library
- SecOC - Secure On-Board Communication.

C. CSM and CAL.

AUTOSAR specifies two crypto modules. The first one is the Crypto Service Manager (CSM) which is located in the service layer of the BSW and provides services for higher-level applications. CSM allows different applications to use the same service to access different cryptographic primitives (cryptographic algorithms). One application may need access to an MD5 digest while another application needs to compute a SHA1 digest. This service of the CSM module can be accessed only locally inside that ECU, while any access between different ECUs needs to be specified and implemented as a separate mechanism, as this is not defined by AUTOSAR [17].

The second crypto module is the Crypto Abstraction Layer (CAL) which is a static library with a very similar function as the CSM. The library is used to provide cryptographic functionality directly by bypassing the run-time environment (RTE). CAL provides C-functions that can be called directly from other software modules such as BSW, software components (SWCs) or even the Complex Drivers layer. The services of the CAL depend on the underlying cryptographic algorithms and are always executed as a call to a function. Because CAL is a library, it is not related to any of the layers in the AUTOSAR architecture [17].

Neither CSM nor CAL defines any actual cryptographic algorithms, instead this is specified by the implementers based on their choice or the customer needs. Because the OEMs choose the cryptographic algorithms to be implemented, it is very important that their security staff are experienced and know which algorithms are secure to use. Both modules provide an input parameter used to select a certain cryptographic algorithm that was requested by a software component or a module.

The following cryptographic functions may be implemented by the CSM or CAL [17]:

- Hash calculations
- Generation and verification of message authentication codes (MAC)
- Random number generation
- Encryption and decryption using symmetrical algorithms
- Encryption and decryption using asymmetrical algorithms
- Generation and verification of digital signatures

5.2 Introduction

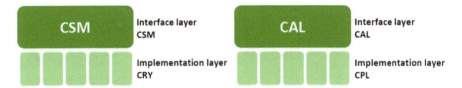

Fig. 5.5 Two layers of crypto modules [11]

- Key management operations

As Fig. 5.5 shows, these two crypto modules are subdivided into two layers: interface layer and implementation layer.

The interface layer is completely standardized by the AUTOSAR while the cryptographic algorithms in the implementation layer are defined by the implementer. The CSM implementation layer is called Cryptographic Primitives Module (CRY) while the implementation layer for CAL is called Cryptographic Primitives Library (CPL). These two modules are used to implement cryptographic algorithms (routines) that will be used by software components (SWCs) in the application layer and modules in the BSW layer. The cryptographic algorithms are implemented using a software library (CSM and CAL) or a cryptographic hardware module (only CSM). Both, CSM and CAL, have to be configured with information about the names of the cryptographic algorithms in the implementation layer (CRY and CPL) and the maximum sizes of keys of the corresponding CRY/CPL module [17].

Figure 5.6 shows the main differences between these two modules. Even though they have similar functionality, CSM is a service while CAL is a library, and they use different communication mechanisms. The existence of both modules is because of historical reasons, as stated in the AUTOSAR documentation.

D. **Secure On-Board Communication (SecOC)**

The SecOC module provides an authentication mechanism for critical data. It is used in all ECUs that require secure communication. This module is specified for the first

	CAL	CSM
Services	25	26
Implementation	Library	System service module
Behaviour	Synchronous	Synchronous / Asynchronous
API	Cal_< Service >(cfgId,ContextBuffer,...)	Csm_< Service >(cfgId,...)
Context buffer	Provided by application	Buffer has to be provided by CRY
Crypto	CPL (Crypto Primitive Library)	CRY (Cryptographic library)
Re-entrance	Re-entrant	Non re-entrant
Usage	Following functions have to be called: ➤ (Csm/Cal)_<service>Start ➤ (Csm/Cal)_<service>Update (at least one time) ➤ (Csm/Cal)_<service>Finish	

Fig. 5.6 Differences between CAL and CSM

time in Release 4.2 of the AUTOSAR standard specification. The module provides a security mechanism that is easy to implement into the existing communication technology, is not resource-heavy, and as such can be used for legacy systems as well [11].

The specification of the module mostly relies on the assumption that symmetric authentication approaches will be used with message authentication code (MAC), but it also supports asymmetric authentication approaches. The symmetric authentication approach is faster and less complex and achieves the same level of security but with smaller authentication keys compared to the asymmetric approach.

Figure 5.7 shows that the SecOC module is integrated on the same level as the protocol data unit router (PduR). On the configuration in Fig. 4.7, the PduR is responsible for the routing of the security-related I-PDUs (information-protocol data unit) to and from the SecOC module. This module then adds or processes the security-relevant information of the I-PDU, and reports the results to the PduR [12].

Figure 5.8 shows the structure of a secured I-PDU that contains authentic I-PDU, freshness values, and the authenticator. Authentic I-PDU is an AUTOSAR I-PDU that requires protection against manipulation and replay attacks. The freshness value and the authenticator ensure the integrity and authenticity of the information.

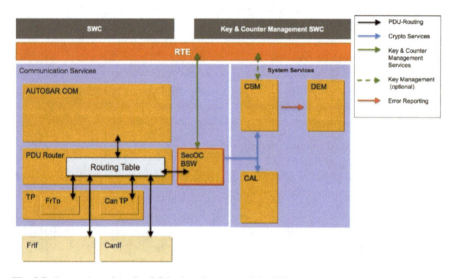

Fig. 5.7 Integration of the SecOC basic software module [12]

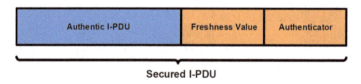

Fig. 5.8 Contents of the secured I-PDU [12]

5.2 Introduction

This security module protects against injection, alteration, and replay of secure I-PDUs. It uses cryptographic algorithms of the CSM or the CAL module and interacts with the RTE layer in order to allow key and counter management. It provides the functionality for verifying the authenticity and freshness of the PDU-based communication between different ECUs inside the vehicle. The authentication and integrity protection ensure that the information sent over the in-vehicular network comes from the right ECU and that the information itself is correct [12].

5.3 Threat Models for the Automotive Domain

In the following, we describe our suggested modifications (based on domain expertise from industry representatives) to make TARA and STRIDE more suitable for automotive threat modeling.

5.3.1 Adaptation of TARA

In order to adapt and apply the TARA method to the automotive industry in general and the connected car in particular, certain modifications were made to the method. The method is intended to be conducted internally inside one single car manufacturer company. The main reason for this is the sensitivity and confidentiality of the information that is needed in order to perform the method successfully. The other reason is the knowledge and the experience of the security experts that work for that specific car manufacturer company. These factors are very important in order to get accurate and reliable results.

For the purpose of this chapter, our goal was to create a more general framework that can be taken and refined within the car companies as described above. For example, the structure of the TAL and MOL library should be adapted, while the CEL library has no strict structure and as such does not require any adaptations. Three sources of information were used. First, we performed an extensive literature survey related to automotive cybersecurity and recent cyberattacks on vehicles. In addition to this, an online survey was also created and filled in by domain experts (seven different companies, including two major car manufacturers) in order to gather valuable input. The process was then followed by individual meetings with a few experts from the industry. The process is further described in a separate report [18], and only the proposed changes are discussed below due to space limitations.

TAL Library. This library lists the names of all the threat agents that are relevant to the automotive industry along with their corresponding attributes. The new adapted version of this library is presented in Sect. 5.4.1. The following changes are made compared to the original TAL library [19] provided by Intel.

- Ten threat agent profiles are removed, and eight new profiles are added: outward sympathizer, hacktivist, cyber vandal, online social hacker, script kiddies, organized crime, cyber terrorist, and car thief. The new agent profiles are based on three sources [13–21] along with research and consultation with domain experts. The "outcome" attribute is modified and now includes the following parameters: acquisition/theft, business advantage, material damage to the vehicle, physical harm to the drivers/passengers, reputation damage, technical advantage, and "15 min of fame."
- Attributes assigned to one threat agent from the original TAL library were slightly modified. The skills and resources level for the sensationalist threat agent were raised to a higher level than in the original model. More explanation is given in Sect. 5.4.1.

MOL Library. This library provides information about threat agent objectives, likely methods they might use, and the impact that their actions would have on the automotive company and the assets in the connected car. The new adapted version of this library is presented in Sect. 5.5.1. The following changes were made to reflect the automotive domain:

- The sections "Acts" and "Limits" are removed and replaced by the "Method" section with the following values: theft of PII and business data, denial of service, intentional manipulation, unauthorized physical access and "unpredictable." The "Limits" section can be found in the TAL library.
- The levels of the "Impact" attribute are replaced with new impact levels: reputation damage, privacy violation, loss of financial assets/car, traffic accidents, and injured passengers. The impact level reveals consequences that an attack would have on the connected car.

5.3.2 Adaptation of STRIDE

The main part of the adaptation of STRIDE is reflected in the template of the MS Threat modeling tool, since the template provides the different elements to create DFD diagrams. Each element is associated with a specific list of threats, and based on the type of interaction between the elements in the DFD diagram, the tool generates a threat report. We use the template developed by the NCC Group [22] with some additions due to the higher abstraction level used here; three new elements are added that represent the underlying architecture that is based on the AUTOSAR standard.

The main reason why this method needs to be adapted is because the method itself was created for the computer industry. The MS Threat modeling tool is also intended to model DFD diagrams for environments such as Windows and Linux operating systems or different applications inside these systems (i.e., web applications, client–server environments). For these reasons, a special template was used, as previously stated, which contains the elements that reflect the vehicle software environment and the associated applications and interfaces. Each of these elements is associated with

5.3 Threat Models for the Automotive Domain

a group of threats that are specifically related to the automotive industry and the connected car. More information is given in Sect. 5.4.2.

5.4 Applying the Adapted Threat Models to the Automotive Domain

5.4.1 TARA

The following sections describe the process of threat modeling using the adapted version of the TARA method. The target of this process is the connected car.

Methodology and Tools. The threat modeling was conducted with the support of domain experts and a project manager from Intel Security in charge of the TARA method. The method is performed in six steps, the goal of which is to find the critical exposures of the connected car. The following is a short description of how these steps were performed:

1. Measure current threat agent risks. By using the online survey that was completed by security experts, the method determined the threat levels of different threat agents (default risk).
2. Distinguish threat agents with elevated risk level. This step was also conducted by using the online survey that determined which threat agents have an elevated risk level when it comes to the connected car as the main target (project risk).
3. Derive primary objectives of those threat agents. By using the survey results and with the support of domain experts, primary motivations and goals of each threat agent were determined and stated in the MOL library.
4. Identify methods likely to manifest. Based on extensive research of previous cyberattacks on vehicles and with the support of domain experts, it was concluded that attacks on vehicles can be classified into five attack methods and can have five impact levels accordingly.
5. Determine the most important collective exposures. The CEL library was created with the support of domain experts, the information gained from the online survey, and extensive research in the field of automotive cybersecurity. The created CEL library relates to the entire automotive industry and not just one specific company; it contains a list of the most exposed areas of the connected car ranked by the level of exposure.
6. Align strategy to target the most significant exposures. The results of this method, stated in Sect. 5.5, can be further used by car manufacturer companies to align their security strategy and focus their resources to the areas of greatest concern.

Threat Agent Library (TAL). The adapted version of the TAL library specifies 19 different threat agents that are relevant for the automotive industry. Each threat agent is described by nine different attributes. The TAL library provides all the information that is needed in order to determine which threat agents present the greatest risk to the connected car. Thus, the TAL library is used by security experts while conducting the first two steps of the TARA method. The results of these steps are given in Sect. 5.5. The following is a list of attributes with a short explanation.

- **Intent** describes whether the agent's intent is to cause harm or not.
- **Access** describes what type of access the agent has to the target: internal (insider) or external (no access to internal data or resources).
- **Outcome** is an attribute that describes the final results of the agent's actions, e.g., actions taken by a threat agent could have business or technical advantage for another competing company by stealing some confidential information.
- **Resource** attribute represents the type of resources the agent has access to, e.g., does the threat agent work alone or in a team with several other threat agents, or it may even have the support of a government implying almost unlimited resources.
- **Skills** attribute describes the level of skill that the agent has.
- **Motivations** are a newly introduced attribute that explains the motivation behind an action conducted by each of the threat agents. Whether it is for personal satisfaction or financial gain, it is important to know because it reveals the reason and the intensity behind the attack.
- **Visibility** describes the extent to which the agent wants to hide or reveal their identity. The victim immediately (overt/covert) knows some attacks while other attacks are hidden (clandestine) and the victim does not know that an attack even took place.
- **Limits** attribute describes the extent to which the agent would go in order to accomplish their goals. Whether the agent would break the law or not is described by this attribute.
- **Objective** describes the primary action the agent will take in order to achieve their goal.

Methods and Objectives Library (MOL). The Methods and Objectives Library (Fig. 5.9) shows the defining motivations (primary cause of their actions) of threat agents, their main goals, and the most likely methods they would employ in order to successfully accomplish their goals. In comparison with the TAL library, where each threat agent has one or more possible motivations, the MOL library just states one main and most likely motivation when it comes to the automotive industry. The decision on which motivation to include in the MOL was based on consultation with domain experts and the online survey. Note that the goal attribute of the MOL library is very similar to the outcome attribute of the TAL library. The difference is, in the MOL library it represents the desired result one wishes to achieve while in the TAL library, it represents consequences of threat agent actions. Based on the research and consultation with the experts, most of the cyberattacks on vehicles today can be summarized with five attack methods. It is difficult to state which specific method each of the threat agents might conduct and without insight into real incident/breach reports the decision was made to categorize the methods on a higher level:

5.4 Applying the Adapted Threat Models to the Automotive Domain

AGENT NAME	ATTACKER Access	Trust: None	Partial Trust	Employee	Administrator	OBJECTIVE Motivation	Goal	METHOD: Theft of PII and Business Data	Denial of Service	Intentional Manipulation	Unauthorized Physical Access	Unpredictable Action	IMPACT: Reputation Damage	Privacy Violated	Loss of Financial Assets / Car	Traffic Accidents	Injured Passengers
Competitor	External	✓				Organizational Gain	Technical advantage	✓					✓	✓			
Car Thief	External	✓				Personal Financial Gain	Acquisition / Theft				✓		✓		✓		
Cyber Terrorist	External	✓				Ideology	Physical harm; Damage			✓						✓	✓
Cyber Vandal	External	✓				Dominance	Personal Satisfaction	✓	✓	✓			✓	✓		✓	
Data Miner	External	✓				Organizational Gain	Technical advantage	✓					✓	✓			
Disgruntled Employee	Internal		✓	✓	✓	Disgruntlement	Reputation Damage	✓	✓				✓		✓		
Government Cyber-warrior	External	✓				Dominance	Physical harm; Damage	✓	✓	✓						✓	✓
Government Spy	Internal		✓	✓	✓	Ideology	Technical advantage	✓	✓	✓	✓			✓		✓	✓
Hacktivist	External	✓				Ideology	Reputation Damage	✓					✓	✓			
Information Partner	Internal		✓			Organizational Gain	Business advantage					✓	✓	✓	✓		
Internal Spy	Internal		✓	✓	✓	Personal Financial Gain	Acquisition / Theft	✓					✓	✓	✓		
Online Social Hacker	External	✓				Personal Financial Gain	Acquisition / Theft	✓						✓			
Organized Crime	External	✓				Organizational Gain	Acquisition / Theft	✓	✓	✓	✓		✓	✓	✓	✓	✓
Outward Sympathizer	Internal		✓	✓	✓	Personal Satisfaction	No Malicious Intent		✓	✓			✓	✓			
Radical Activist	External	✓				Ideology	Material Damage	✓	✓	✓			✓	✓		✓	
Reckless Employee	Internal		✓	✓	✓	Accidental / Mistake	No Malicious Intent					✓	✓	✓			
Script Kiddies	External	✓				Personal Satisfaction	"15 Minutes of Fame"	✓	✓	✓			✓	✓	✓		
Sensationalist	External	✓				Notoriety	"15 Minutes of Fame"	✓					✓	✓			
Untrained Employee	Internal		✓	✓	✓	Accidental / Mistake	No Malicious Intent					✓	✓	✓			

Fig. 5.9 MOL Library

- **Theft of PII and business data**. The threat agent can employ a variety of methods for stealing data personally identifiable information (PII) from the vehicle or from the car manufacturer company. Business data can also include technical data about the company's products, production processes, and technologies they are developing.
- **Denial of Service**. The method that has the biggest potential to be used in the automotive industry is ransomware. The attacker would infect the vehicle with ransomware by exploiting one of the attack surfaces. This would prevent the driver from using the car until the ransom is paid.
- **Intentional manipulation**. This method refers to any type of attack that gives the attacker access to the control functions of the vehicle such as the steering wheel, brakes, and engine. Having access to these functions can allow the attacker to cause traffic accidents, traffic jams or even to cause serious or deadly injuries.
- **Unauthorized physical access**. Different methods for carjacking are the main ones that are covered by this category, but there can also be other reasons for unauthorized access to the vehicle. Attacker could inject malware over the USB interface or the OBD port for later remote access.
- **Unpredictable.** The main purpose of this method type is to reflect the methods of the employees (threat agents) and the information partner. These threat agents do not have a malicious intent but rather through mistakes and accidental actions create a harmful situation for the company.

The final attribute of the MOL library is the impact of the actions taken by threat agents, which can refer to the car manufacturer company, the vehicle, or the PII information stored in the vehicle memory.

Common Exposure Library (CEL). The CEL library does not have a standardized format because it contains confidential information and is derived by each organization. The library maps existing security controls to each of the identified exposures and then compares those security controls with the list of recommended security controls. By doing so, the library provides insight into the residual risk with respect to the existing security control compared to the ones recommended by security standards. This information is very sensitive, confidential, and specific to each car manufacturer which is why it was not included in our Common Exposure Library. Therefore, the CEL library presented here is not complete, but still gives important information about the greatest exposures in the automotive industry. The list of exposures and the rankings is based on an extensive literature survey and consultation with security experts from seven different companies, including two major car manufacturers. Figure 5.10 shows the CEL library with rankings from the highest exposure (OBD II port) to the lowest exposure (CD/DVD player).

The library has three attributes that describe each of the listed exposures:

- **Level**. It describes the level of vulnerability. It has three levels: high, medium, and low.
- **Type of Access**. It describes the most likely type of access needed in order to successfully perform an attack, either physical access or wireless access is considered the most likely.

Level	Exposures	TYPE OF ACCESS		IMPACT POTENTIAL		
		Physical access	Wireless access	Safety	Data Privacy	Car-jacking
HIGH	OBD II port	✓		✓		
HIGH	Wi-Fi		✓	✓		
HIGH	Cellular connection (3G/4G)		✓	✓		
HIGH	Over-the-air update		✓	✓		
HIGH	Infotainment System		✓	✓		
HIGH	Smart-phone	✓		✓		
MEDIUM	Bluetooth		✓	✓		
MEDIUM	Remote Link Type App		✓	✓		
MEDIUM	KeyFobs and Immobilizers		✓			✓
MEDIUM	USB	✓		✓		
MEDIUM	ADAS System		✓	✓		
MEDIUM	DSRC-based receiver (V2X)		✓	✓		
LOW	DAB Radio		✓	✓		
LOW	TPMS		✓		✓	
LOW	GPS		✓		✓	
LOW	eCall		✓	✓		
LOW	EV Charging port	✓		✓		
LOW	CD/DVD player	✓		✓		

Fig. 5.10 CEL Library

5.4 Applying the Adapted Threat Models to the Automotive Domain

- **Impact Potential**. The impact describes the potential impact if the attack surface is successfully exploited. It has the following three parameters:

1. Safety is the highest impact level, which means if the attack surface is exploited, the possibility of affecting the safety of the passengers is high
2. Data privacy level refers to attack surfaces which if exploited could lead to privacy violations, and
3. Carjacking describes an attack surface that directly relates to unauthorized physical access to the vehicle (car theft).

5.4.2 STRIDE

To demonstrate the STRIDE analysis, we choose a specific software application: the interior lights of the car. The main reason for choosing this application is that the interior lights application is available as an AUTOSAR application, both for simulation and as a runnable on real hardware. Furthermore, the interior lights application implements information flow from the ECU level up to the application level, which is of interest for STRIDE analysis. We can then demonstrate the modeling (this section) and the results (with hardware validation, discussed in Sect. 5.5.2).

Methodology and tools. The STRIDE analysis is performed on the AUTOSAR platform provided by the company Arccore.

First, analysis of the interior lights application was conducted, so that the DFD diagrams could be derived. These were then created by the Microsoft Threat modeling tool by using the NCC Group template. These steps were conducted with the support of domain experts from Arccore and the NCC Group. Finally, the threat report was generated and examined in order to exclude false positives. Additional validation was conducted by testing some of the found threats in a simulated vehicle environment, more information on this is given in Sect. 5.5.2.

The Interior Light Application. The application consists of seven different software components (SWCs), such as the light actuator and the door sensor SWC, each providing a specific function for the interior light application. The application receives input data from the sensors (door sensor SWC) that notify the application if the vehicle door is open/closed and if

the car trunk is open/closed. After analyzing the input data from the sensors, the application sends signals to the actuators (light actuator SWC) that control the interior light of the vehicle and notifies them if the lights should be turned on/off. The information is exchanged using the CAN network.

Data Flow Diagrams. This section introduces the DFD diagram (Fig. 5.11) that represents the interior light application. The DFD diagram is based on two previously created diagrams, one representing AUTOSAR communication services and the other representing the AUTOSAR I/O services. In Fig. 5.7, these diagrams are summarized in two nodes, "AUTOSAR COM Layer" and "AUTOSAR I/O Layer," respectively.

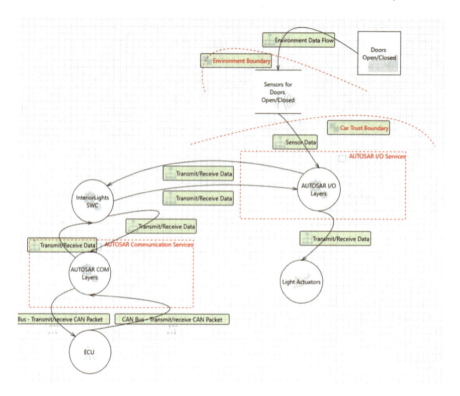

Fig. 5.11 DFD created with the MS Threat modeling tool and the NCC Group template

AUTOSAR communication services are in charge of transferring messages from other ECUs on the CAN network and messages that could come from external and potentially malicious users. AUTOSAR I/O services provide access to sensors and the actuators that the interior light application needs in order to function properly. The entire DFD diagram is not included, as most threats would be found on trust boundaries and interfaces to external actors.

5.5 Results

5.5.1 TARA

Results of the TARA method are reflected in the three libraries that were created during the threat modeling process and the risk comparison shown in Fig. 5.12. It is especially important to analyze the CEL library as it contains a list of the most exposed areas and the graph as it identifies the threat agents that pose the greatest risk. The CEL library identified the OBD port as the interface that is exposed the

5.5 Results

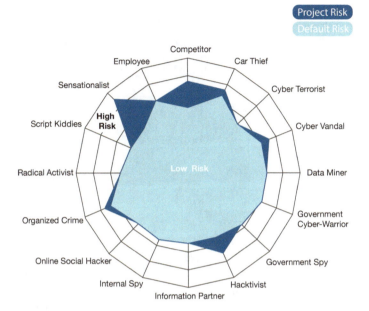

Fig. 5.12 Risk comparison of different threat agent profiles

most followed by the Wi-Fi connection in the second place. The rest of the exposed areas are then listed in descending order. Figure 4 is based on the survey results and the support of domain experts. It shows risk levels for each threat agent when the connected car is taken as the target. The default risk, as stated in the figure, represents the general risk to different IT services while the project risk represents the connected car. The method identified six threat agents that have an elevated risk

level, where the sensationalist was identified as the one posing the greatest risk. This threat agent refers to people who wish to attract public attention by employing any method for notoriety just to get their "15 min of fame."

5.5.2 STRIDE

After the MS Threat modeling tool analyzed the DFD diagram for the Interior light application, it generated a threat report with 74 different potential threats, where at least one threat from each STRIDE category was found. It also found 17 threats that were not applicable to the application.

To ensure that the results of the threat modeling process are credible and should be further analyzed by experts, we verified the found threats with an actual hardware implementation of the application. Testing was performed on an AUTOSAR hardware board (with the interior light application) connected to a small CAN network and a computer to analyze the communication and the exchanged packets (Fig. 5.13). By

Fig. 5.13 DFD created with the MS Threat modeling tool and the NCC Group template

conducting this validation process, we could investigate whether the threats generated by the threat modeling tool are applicable to the actual AUTOSAR software application, and as such, to a real vehicle system.

The validation process was conducted successfully, and the threats discovered by the threat modeling process were confirmed. Hence, the adapted STRIDE method can be applied to other systems in the automotive domain and as such become a valuable tool for automotive security experts.

5.5.3 Related Work

Very little research has been found regarding threat modeling of the connected car. Two of the first papers that discuss security analysis of modern vehicles were published in 2010 and 2011 by researchers Koscher et al. [23] and Checkoway et al. [24]. They conducted a series of experiments and road tests on a modern vehicle, and among other things found that it was possible to manipulate the vehicle by injecting false messages on the CAN network. They also analyzed the external attack surface of the vehicle and managed to exploit some of the connection points, i.e., Bluetooth function and the vehicle's cellular connection used for the telematics system. However, the team that conducted the research did not use any specific methods for analyzing the vehicle's attack surface. The work by Wolf and Scheibel [25] from 2012 was one of the first that gave a new risk analysis method tailored to the needs of the automotive industry. Their method considers two factors: potential damage and the probability of a successful cyberattack. The goal of the method is to avoid over securing or under securing different parts of the vehicle which in return decreases costs. Furthermore, in 2014 the US National Highway Traffic Safety Administration (NHTSA) [26] presents a composite modeling approach of cybersecurity threats to the connected vehicles. They created threat models and threat reports of different types of possible threats to vehicles along with a list of potential attacks.

Threat analysis methods have also been presented by the following authors. The first, by Mundhenk et al. [27], presents a new method for security analysis. This method uses continuous-time Markov chains (CTMC) to model the architecture at the system level. Afterward, the model is analyzed for confidentiality, integrity, and availability by using probabilistic model checking. Another approach to combine safety and security risk analysis was proposed by Machera et al. [28]. The authors combined the automotive hazard analysis and risk assessment (HARA) with the security domain threat modeling STRIDE. The resulting method was named Safety-Aware Hazard Analysis and Risk Assessment (SAHARA). This method is used to determine the impact of security threats to the safety concepts in the vehicle at system level. The most recent paper, by Islam et al. [29], gives some more information about the process of risk assessment for the automotive embedded systems. The authors combine the threat analysis with risk assessment in order to determine a security level that indicates what level of protection a certain part of the system needs. The described papers introduced some new threat modeling methods; however, the application of the TARA method and the usage of MS Threat modeling tool with the NCC Group template has not been done before. This chapter tries to fill this gap and present these methods and tools to the automotive industry.

5.5.4 Discussion and Future Work

The TARA method is rather new with little supporting documentation except what is published by Intel Security [7]. For that reason, additional work had to be conducted in order to successfully adapt and apply the method to the automotive industry.

The method ranked the sensationalist and the OBD port as having the highest risk in their respective category. The former mirrors well the majority of the attacks that have been widely documented and discussed. These were conducted by different researchers and experts that wanted to show how insecure the vehicles really are. The final goal of the researchers was maybe

not to get famous and hit the headlines of all news portals in the world, but it was definitely the outcome of their research and as such has to be taken into consideration. The OBD port is shown to have the highest risk potential, and even though it requires physical access there are some cases where it can be exploited remotely [6]. It is also the oldest interface in the CEL library and accessing the OBD port gives the attacker almost full access to the CAN network.

Unlike the TARA method, STRIDE has been used before in the automotive industry. However, the method in this chapter was conducted by using the MS Threat modeling tool along with the template designed for the automotive industry, and this was not done before to the best of our knowledge. The template used with the MS

Threat modeling tool has shown to be very useful and adaptable, where further work may allow this template to be used on even more low-level software applications.

The results generated by the tool described in this chapter are maybe not completely comprehensive but they clearly show the extent of vulnerabilities of an AUTOSAR-based software application. Even though the application in question, the vehicle's interior lights, does not seem like something worth analyzing, one can just imagine driving down the highway in the middle of the night when suddenly the lights in the car start going on/off every second—it could distract the driver or even cause an accident.

5.6 Conclusion

In this chapter, we adapted two threat modeling methods from the computer industry in order to better fit the needs of the automotive industry. The next step was to apply these methods to the connected car and the underlying software architecture, which in turn generated valuable results that were carefully validated. The entire work was done with the support of domain experts from different companies that have extensive knowledge in this field. TARA was used to provide a high-level overview of threats in the area of connected vehicles while STRIDE was used to evaluate a specific functionality of the vehicle.

The three libraries created by the TARA method and the template used by the STRIDE method would be a good starting point for any future application. The research described in this chapter, including the actual validation of STRIDE results on real hardware, demonstrates the usefulness of these methods and domain experts should be able to include them in their tool set for future analysis.

Finally, it is important to learn from the mistakes made by the computer industry, but it is also vital to recognize which threat modeling methods and which security mechanisms from the computer industry can be applied to the automotive industry. We need to use the existing technology and experience, adapt it to fit the automotive industry, and apply it to secure the vehicles on our roads. As the vehicles are more connected with the introduction of V2X technology and the autonomous driving, more threat modeling methods will be needed. The only way to build a secure connected car is to incorporate security from the start and not as an afterthought.

References

1. R. Currie, M. Santander, "Developments in car hacking". in *SANS Institute InfoSec Reading Room* (2015)
2. J. Greenough, "The Connected-Car Report: The Transformation of the Automobile". in *Bussines Insider Intelligence* (2016)
3. C. Miller, C. Valasek, "Remote Exploitation of an Unaltered Passenger Vehicle". in *Black Hat USA* (2015)

References

4. International Business Times. *Tesla Model S hacked: Researchers discover six security flaws in popular electric car.* Accessed 2016-10-03. http://www.ibtimes.co.uk/tesla-model-s-hacked-researchers-discover-six-security-flaws-popular-electric-car-1514352
5. International Business Times. *Hacker takes control of Nissan electric vehicle from other side of the world through Leaf app.* Accessed 2016-10-03. http://www.ibtimes.co.uk/hacker-takes-control-nissan-electric-vehicle-other-side-world-through-leaf-app-1545808
6. International Business Times. *Hackers disable Corvette brakes by texting dongle meant to lower insurance risk.* Accessed 2016-10-03. http://www.ibtimes.co.uk/hackers-disable-corvette-brakes-by-texting-dongle-meant-lower-insurance-risk-1515125
7. M. Rosenquist, Prioritizing Information Security Risks with Threat Agent Risk Assessment 2009. https://itpeernetwork.intel.com/whitepaper-prioritizing-information-security-risks-with-threat-agent-risk-assessment/. (2009)
8. OWASP. Threat Risk Modeling. https://www.owasp.org/. (2016)
9. Vector Corp. http://www.elearning.vector.com. Accessed 2016-10-03. (2015)
10. AUTOSAR. Utilization of Crypto Services - AUTOSAR Release 4.2.2. Accessed 2016-10-03. http://www.autosar.org/standards/classic-platform/release-42/
11. AUTOSAR. www.autosar.org. Last Accessed on 2016-10-03
12. AUTOSAR. Specification of Module Secure Onboard Communication AUTOSAR Release 4.2.2, http://www.autosar.org/standards/classic-platform/release-42/. Last Accessed on 2016-10-03
13. M. Wille. Automotive security—an overview of standardization in AUTOSAR. VDI/VW-Gemeinschaftstagung Automotive Security, 2015. Wolfsburg
14. ENISA. Threat Landscape 2015. https://www.enisa.europa.eu/publications/etl2015. (2015)
15. C. Bernardeschi, G. Din, Security modeling and automatic code generation in AUTOSAR. Università di Pisa Dipartimento di Ingegneria dell'Informazione Corso di Laurea Magistrale in Computer Engineering, 2016. https://etd.adm.unipi.it/theses/available/etd-04042016-183740/unrestricted/TesiAUTOSAR.pdf
16. W. Trumler, M. Helbig, A. Pietzowski, B. Satzger, Self-configuration and self-healing in AUTOSAR. Asia Pacific Automotive Engineering Conference, 2007
17. AUTOSAR. Utilization of crypto services—AUTOSAR release 4.2.2, http://www.autosar.org/standards/classic-platform/release-42/. Last Accessed on 2016-10-03
18. Adi Karahasanovic. Automotive Cyber Security—Threat modeling of the AUTOSAR standard. http://studentarbeten.chalmers.se/publication/247979-automotive-cyber-security. (2016)
19. T. Casey, "Threat Agent Library Helps Identify Information Security Risks". in *White Paper Intel Information Technology* (2007)
20. D. Houlding, T. Casey, M. Rosenquist, "Improving Health-care Risk Assessments to Maximize Security Budgets". in *Intel White paper* (2012)
21. T. Casey, "A field guide to insider threat". in *Intel White paper* (2015)
22. NCC Group. The Automotive Threat Modeling Template. Accessed 2016-11-04. https://www.nccgroup.trust/uk/about-us/newsroom-and-events/blogs/2016july/the-automotive-threat-modeling-template/
23. K. Koscher, A. Czeskis, F. Roesner, S. Patel, T. Kohno, S. Checkoway, et al. "Experimental Security Analysis of a Modern Automobile". in: *2010 IEEE Symposium on Security and Privacy* (2010)
24. S. Checkoway, D. McCoy, B. Kantor, D. Anderson, H. Shacham, S. Savage, "Comprehensive Experimental Analyses of Automotive Attack Surfaces". in *SEC'11 Proceedings of the 20th USENIX conference on Security* pp. 6–6 (2011)
25. M. Wolf, M. Scheibel, "A Systematic Approach to a Quantified Security Risk Analysis for Vehicular IT Systems". in *Automotive—Safety and Security 2012 Conference, Karlsruhe* (2012)
26. C. McCarthy, K. Harnett, A. Carter, NHTSA—Characterization of Potential Security Threats in Modern Automobiles A Composite Modeling Approach. https://trid.trb.org/view.aspx?id=1329315. (2014)
27. P. Mundhenk, S. Steinhorst, M. Lukasiewycz, S.A. Fahmy, S. Chakraborty, "Security Analysis of Automotive Architectures using Probabilistic Model Checking". in *DAC'15 Proceedings of the 52nd Annual Design Automation Conference* (2015)

28. G. Machera, E. Armengauda, E. Brennerb, C. Kreinerb, "Threat and Risk Assessment Methodologies in the Automotive Domain". in *The 1st Workshop on Safety and Security Assurance for Critical Infrastructures Protection (S4CIP)* (2016)
29. M.M. Islam, A. Lautenbach, C. Sandberg, T. Olovsson, "A Risk Assessment Framework for Automotive Embedded Systems". in *Proceedings of the 2nd ACM International Workshop on Cyber-Physical System Security* (2016)

Chapter 6
Inter-Vehicle Communication and Cyber Security

6.1 Overview

In this chapter, we focus on the connected vehicle technology and describe in detail about the different types of connected vehicle technology and its security issues. The intelligent and autonomous vehicle (IAV) technologies are expected to revolutionize the intelligent transportation system internationally. The IAV can be categorized into autonomous vehicles (or self-driving vehicles) and connected vehicles. The inter-communication between vehicles is known as connected vehicles in the USA [1] and Cooperative Intelligent Transport Systems (C-ITS) in Europe [2]. The autonomous vehicles are based on a mixture of varieties of sensors and diverse technologies to achieve the desired autonomous level, while connected vehicle technologies are based on vehicular ad hoc networks (VANETs). The connected vehicles help the vehicles to make intelligent decisions such as establish exchanged messages between the neighbor vehicles and the RSU. This helps in planning the future travel route. This makes the IAV to be safer as well as improves the traffic flow with very low probability of accidents. The vehicle-to-everything (V2X) is the main communication technology for future VANETs that helps vehicles to obtain a wide range of road information in real time that significantly improves driving safety, traffic efficiency as well as provides infotainment services. However, connected vehicles are based on radio access technologies like 802.11p. The V2X applications cannot handle the complexity and challenges due to huge number of generated data from cooperative sensing, high density of platooning vehicles, and multimedia demands. VANETs have dynamic topology and fast evolving communication demands such as ultra-low latency and ultra-high reliability. An efficient radio access technology, core network functionalities, and edge cloud services are required to overcome the V2X demands. Due to the proven history, stability, and advancement of cellular technologies in telecommunication, the cellular technologies can tackle the existing issues and demands of V2X communication in VANET. The third partnership project (3GPP) releases specification in Rel. 14 and Rel. 15 for cellular V2X (C-V2X) and working on further developing to 5G technology for V2X communication in Rel. 16

[3]. The new radio technology for 5G can be a game-changer as it can complement or replace the 802.11p in V2X communication. The objective of this chapter is to provide detail information in the race toward enabling V2X communication based on advanced radio technology such as cellular 5G technology.

6.2 Connected Vehicles

In the USA, the vehicles that are wirelessly connected with other vehicles, infrastructures, and clouds are popularly known as connected vehicles (CV) technologies [1]. The connected vehicles technology is based on ad hoc networks to communicate with other nodes known as VANET. VANETs enable intelligent vehicles to transmit vehicle-related safety information to prevent accidents. In VANET, Wireless Access in Vehicular Environments (WAVE) protocol provides the basic radio standard for dedicated short-range communication (DSRC) operating in the 5.9 GHz frequency band, which is based on the IEEE 802.11p standard [4]. In Europe, European Telecommunication Standards Institute (ETSI) has started equivalent standard similar to DSRC, which is the only commercially available short-range V2X technology. It is known ETSI ITS G5 (ITSC) standard [5] to support vehicle-to-everything (V2X) communication. The ITSC standard is created based on some revisions in IEEE 802.11p standards with European Union prerequisites as ETSI EN 302 663 [5]. This enables the vehicles to communicate with infrastructure using DSRC within 500 m [6].

6.2.1 VANET Technology Overview

The connected vehicles technology or C-ITS is based on VANET for communicating safety and non-safety messages. Due to the ITS technological advancement, the connected vehicles have gained a lot of attention from industry and academia that could potentially help the safe driving, traffic flow situation such as congestion, accident, road construction, and multimedia entertainment in vehicles. The basic aim of connected vehicles or C-ITS is to enhance road safety and traffic flow efficiency with reduced global pollution and effective environmental impact. This can be achieved by enabling IAV and roadside infrastructures to communicate and exchange vehicle-related safety messages that consist of road hazards, speed, location, size, and direction information. The connected vehicle technology or C-ITS is based on VANET for communicating safety and non-safety messages. In VANET, the vehicle can communicate with neighbor vehicles, infrastructure, central networks, pedestrians, cyclist, etc., based on 802.11p radio technology in unlicensed 5.9 GHz band. The wireless communication in VANET can be achieved by

6.2 Connected Vehicles

(i) Vehicle-to-infrastructure (V2I) communications in the infrastructure domain. In V2I, the vehicles connect to road side units (RSUs) within its communication range. The RSU can be implemented in either an eNodeB (eNB) or stand-alone traffic light posts.
(ii) Vehicle-to-network (V2N) communications to connect with remote servers, the evolved packet switching and cloud-based services than can be connected through cellular networks.
(iii) Vehicle-to-vehicle (V2V) communications in ad hoc domain to connect with other neighbor vehicles in close proximity supporting distributed localized interaction among the neighbor vehicles without RSUs.
(iv) Vehicle-to-pedestrian (V2P) to connect with the vulnerable pedestrians or bikers on the road in ad hoc mode like V2V communication.

The term V2X collectively refers to communication among different entities based on V2I, V2V, V2N, V2P, etc., for both safety and non-safety applications as mentioned in 3GPP document [7].

Besides this, there are other V2V communication protocols based on wireless network technology in vehicular communication. Some of the short-range technologies such as Zigbee, Bluetooth, and Near-Field Communication (NFC) can be used for communication between two nodes that are close to each other. Due to the shortage of radio frequency spectrum, there is another alternative technology based on visible light technology (using LED lights and car taillights) such as visible light communication (VLC) and light fidelity (Li-Fi). The VLC based on LED technology has long lifespan, lower power consumption, high efficacy, and is weather tolerant; however, line-of-sight is required for efficient communication. There are only few researches carried out based on VLC technology for vehicular communication.

6.2.2 Types of Communications Technology in Connected Vehicle

The advancement in vehicle connectivity includes different types of communication technologies. At present, there are two underlying technologies that enable V2X communication, namely IEEE 802.11p and cellular technologies. We begin with the V2X communication techniques and then explain how V2X based on 802.11p and cellular V2X (C-V2X) technologies empower these techniques to contribute ITS applications. Figure 6.1 shows two types of wireless technologies used in connected vehicles. The detailed description of DSRC technology and WAVE communication and its standard suites and applications can be found in Chap. 2. Later in this chapter, we will discuss the security protocols in WAVE communication.

In general, the V2X communication can be direct communication, fully network-assisted communication, and hybrid communication as shown in Fig. 6.2. In direct communication, vehicular nodes communicate with adjacent vehicles based on peer-to-peer protocol. In fully network-assisted communication, the communication has

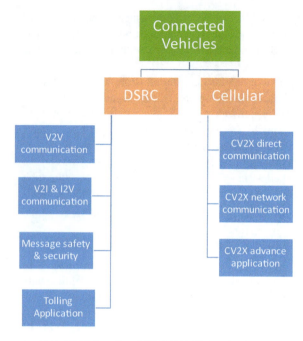

Fig. 6.1 Connected vehicle (V2X) based on DSRC (802.11p) and cellular technology

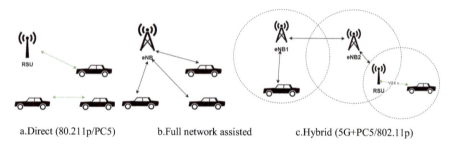

a. Direct (80.211p/PC5)   b. Full network assisted   c. Hybrid (5G+PC5/802.11p)

Fig. 6.2 Different types of V2V and V2I scenarios

to go through the RSU or base station (BS). In hybrid communication, the vehicular node communicates with each other through the combination of direct communication and network-assisted communication using 802.11p and cellular networks such as RSU and BS.

6.3 State-of-the-Art Technologies in VANET

6.3.1 DSRC-Based V2X

The 5.9 GHz DSRC-based V2X supports cooperative awareness applications such as vehicle warning, emergency brake light, and vehicle platooning. However, these applications are suitable only for low density of vehicles. In reality, there are hundreds of thousands of vehicles driving on the road and the V2X applications require very high throughput, high bandwidth, and very low latency in a congested scenario. In critical and emergency warning messages, latency plays a very important role in traffic safety to prevent the vehicle from the accidents. This technology has limitation to fulfill all the requirements of future V2X applications due to the restriction in physical layer (i.e., radio technology) and lack of collision and interference management. This is one of the reasons for delay in implementation of V2X realization in the vehicles. In addition to this, the IAV communication requirements and applications are increasing rapidly while the network configuration, latency, poor scalability, mobility, enormous scale network deployment, security, and achievable data rates of DSRC-based vehicular communication cannot get closer to the ever-growing need of these applications. There are several challenges with DSRC, and it cannot evolve significantly to keep up with the increasing advanced and progressive use cases of V2X technology. The infrastructure mode of DSRC requires RSUs that take huge amount of time and money to be deployed globally. DSRC is an asynchronous system based on CSMA/ CA protocol so there are several inadequacies at the physical layer resulting in reduced performance. In addition, physical/MAC layers in DSRC have limited evolutionary path for further improvements considering reliability, robustness, and range. On the other hand, there is a parallel research going on cellular technology for V2X communication as a candidate technology.

6.3.2 Cellular-Based V2X

The third-generation partnership project (3GPP) gives highest priority for modifications of radio access suitable for V2X communications [8]. The cellular technology has already proven its successful implementation in the last few decades, and the 3GPP inherits the benefits of cellular technology and combines the application of VANET. Based on Rel. 14 and Rel. 15 specification, the 3GPP announced the cellular vehicle-to-everything (C-V2X). In 2016, the 3GPP standardized the use of LTE as cellular networks in the licensed band for V2X communication in Rel. 14, which is known as LTE-V2X [9]. The LTE-V2X is based on already deployed cellular networks that overcomes the limitation of 802.11p-based V2X communication, i.e., it provides very high bandwidth, very low latency, and wide coverage. In C-V2X, it extends the cellular device-to-device (D2D) communications specification by introducing two more operational modes dedicated to V2V communications for vehicles

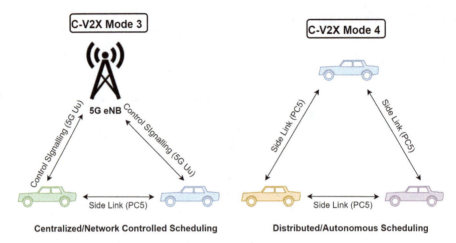

Fig. 6.3 C-V2X Mode 3 and Mode 4 scheduling techniques

[10]. The two communication modes are Mode 3 and Mode 4 as shown in Fig. 6.3. The Mode 3, i.e., scheduled mode operates only in the presence of base stations or eNB. The resource allocations are carried out in a centralized manner by the cellular network. The Mode 4, i.e., autonomous mode operates independently in absence of eNB supporting direct communication utilizing side link radio interface called PC5 using 5.9 GHz frequency band that is similar to DSRC. It can communicate with other neighbor vehicles in a distributed manner without relying on the core cellular network connectivity. The Mode 3 in C-V2X has some issues related to scheduled operation. The vehicular nodes should always be connected with the LTE network, which is very difficult in highway scenarios with high mobility. In this situation, the vehicles frequently disconnect with the eNB due to mobility management problem and hamper the V2X transmission. Hence, the Mode 3 cannot carry out the resource allocation. On the other hand, Mode 4 works considerably better for C-V2X communication.

6.3.2.1 Advancement in Cellular V2X

As the 3GPP came into picture for C-V2X technology, they are working rapidly on the advancement of the C-V2X by releasing several specification standards, to enhance the technology further. At the beginning of DSRC technology, cellular technology-based 3G technology was also in existence but they were insufficient to support the strict requirements essential for V2X communications. With the development of cellular technology, 4G or LTE networks were introduced with high bandwidth, low latency, high throughput, and high reliability. The LTE's infrastructure networks are already widespread with the deployment of large number of eNB that support huge communication traffic between the mobile users.

6.3 State-of-the-Art Technologies in VANET

(a) **LTE-V**: In vehicular networks, LTE for vehicles (LTE-V) was introduced as an alternative technology besides DSRC for intelligent transportation. Telecom operators and automotive industries have accepted LTE-V for vehicular communication. It claims to provide low cost, rapid development, and implementation by using the existing cellular base stations making the urban transportation system more efficient and manageable. After sometime, the telecom company introduced LTE Advanced (LTE-A) networks with higher capacity compared to LTE. Several vehicles use on-board system to connect to the cellular network using LTE/LTE-A for telematics and GPS systems. Some of the applications are infotainment, fleet management, and infotainment. However, LTE and LTE-A still cannot support the ultra-low latency, ultra-high reliability, and ultra-high bandwidth requirement for demanding V2X applications. They cannot guarantee ultra-low latency for time-critical messages like accident messages and cooperative platooning.

(b) **5G C-V2X**: The 3GPP standardization has released a dedicated set of criteria for supporting V2X applications in the future cellular networks based on the 5G technology [11] that is known as cellular V2X (5G C-V2X). As 5G C-V2X is compatible with 4G and 5G, the 3GPP is working toward improving the 5G architecture in Rel. 16 [3]. The 5G technology that supports the requirements of V2X application is still in research, and the automotive and telecom industries are positive and working toward the achievement of 5G C-V2X by 2020. The 5G cellular architecture for V2X is expected to meet the higher performance demands, ultra-low latency, high mobility accommodation, seamless, and reliable exchange of messages and other critical requirements of V2X applications. The global industries from automotive, telecommunication, and technology industries created an association known as 5G Automotive Association (5GAA). It is an international, cross-industry consortium, which defines the C-V2X technology and its evolution toward 5G. Before 5GAA, different industries in different parts of the globe have their own definition, terminology, interest, and use cases of V2X that make it difficult to communicate in one common language. The 5GAA solves this issue by forming a common global language among various industries for V2X underneath one umbrella. The 5GAA objective is to develop, test, initiate standards, promote solutions for C-V2X as well as endorse 5G communication solutions that help to accelerate the global market penetration, accessibility and initiate the C-V2X standardization. The 5GAA objective is to show long-term promise of C-V2X functionality and offers superior performance and ever-increasing capabilities based on 3GPP cellular technology, which is better than IEEE 802.11p [12]. The 5GAA leverages the upper layer standards of IEEE, SAE, and ETSI-ITS. The consortium of automotive industries is continually testing and improving the C-V2X in the ITS sector. The 5GAA foresees the advantages of adopting 3GPP technical specification for C-V2X without any additional requirements for infrastructure network as compared with 802.11p. The main advantage of C-V2X is that it can leverage the cellular ecosystem, high-density support and enhance range, high reliability. In addition, there is opportunity for evolution toward 5G technology.

Table 6.1 shows the comparison between the C-V2X and DSRC that shows superior technical advantages of C-V2X over DSRC in terms of technology, spectral efficiency, connectivity latency and shows higher level of safety to more vehicle users [12, 13]. According to 5GAA, the C-V2X based on device-to-device communication operating in 5.9 GHz frequency, which is a primary common spectrum, can be used for basic safety applications around the globe. This is one of the significant advantages of C-V2X over DSRC for direct communication to operate in an interoperable manner.

(c) **5G NR C-V2X**: The 3GPP focused on enhancing the LTE features on PC5 direct communication and LTE-Uu interfaces before Rel.15. In 2015, the 3GPP started to work on new radio (NR) standardization activity for 5G system as a first phase in Rel.15. The 3GPP released its first full specification of 5G NR in Rel. 15 in June 2018. The 3GPP continued to enhance the C-V2X and accelerated the work on standardization as well as early trial and developments in 5G NR in Rel. 16 in 2019. Identical to the 3G/4G evolution, 5G NR evolution is the part of continuous wireless broadband development process to meet the demands of 5G technology. The core objective of 5G NR evolution is to enhance mobile broadband similar to wired technology like optical fiber with low cost, high capacity,

Table 6.1 Detailed comparison of DSRC-based and cellular-based V2X communication

Components	DSRC V2X	Cellular V2X
Technology	Wi-Fi	LTE/5G
Modulation	OFDM	SC-FDM
Concurrent Tx	No	Yes
Cellular connectivity	Hybrid mode, i.e., connect with cellular network for non-safety services	Hybrid model, i.e., connect with cellular network for non-safety services
Transmission scheduling	CSMA: No predetermined Tx slots and transmit when there is no ongoing reception	Collisions are not sensed. Slow response to changing environment
Time synchronization	Loose asynchronous	Very tight synchronous requirements
Line coding	Convolution code	Turbo code
Deployment	From 2017. Commercialization in 2019	Mass market deployment in China from late 2020
Future guideline	Backward compatible and interoperable upgrade to 802.11p, i.e., 802.11bd	C-V2X Rel.16 is based on 5G NR technology and operates in different channel than previous releases
Latency	Low latency for V2V communication	Round trip latency less than 1 ms, Slight delay due to centralized communication
Range	Good for short radio range	Good for long-range communication

6.3 State-of-the-Art Technologies in VANET

high reliability, very low latency, etc., that will offer several faster, more reliable, and time-critical services. According to International Telecommunication Union (ITU), the major applications of 5G NR technology are ultra-reliable and low latency communications (URLLC), enhanced mobile broadband (eMBB), and massive machine-type communicator (mTC) [14]. The ultra-low latency provided by URLLC, i.e., less than 1 ms is suitable for C-V2X applications. Few other technologies such as massive multiple-input multiple-output (mMIMO), millimeter waves (mmWave) [15], and beamforming assist to fulfill the requirements of 5G NR technology. Figure 6.4 shows the evolution of new 5G NR technologies as well as expansion of 5G ecosystem beyond Rel. 17 in terms of 5G NR CV2X technology for autonomous vehicles.

The standardization of new radio technology for 5G C-V2X in Rel. 16 provides adequate requirements for today's automotive industry as well as future autonomous driving. The 5GNR C-V2X (5NC) brings new opportunities and capabilities for future autonomous vehicles while maintaining backward compatibility at the same time. Some of the new capabilities to C-V2X communications are wideband transmission for accurate ranging and positioning of the vehicles, new design for ultra-low latency slot structure, scalable OFDM technology, supporting massive MIMO, etc. The 5GNR C-V2X provides unifying connectivity platform for future autonomous vehicles such as massive IoT, mission-critical services, and enhanced wireless broadband. It allows wideband carrier support providing high throughput and low latency.

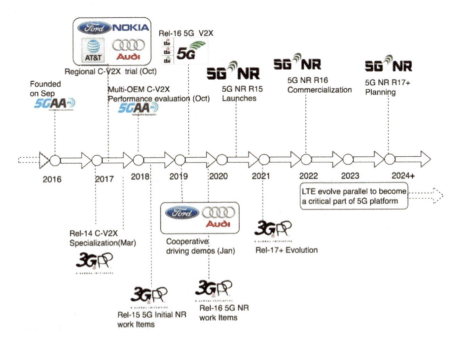

Fig. 6.4 Evolution of C-V2X toward 5G technology for autonomous vehicles

This enables multiple functions such as sharing and exchange of processed or raw information collected from multi-sensors that have been installed in the intelligent and autonomous vehicles, sharing of planned route or intention of the trajectory or next movement. The wideband supports high definition of local dynamic maps, i.e., real-time 3D HD maps, street view information from high-quality street cameras at the interactions, etc.

According to Maxime Flament, who is the CTO at 5GAA expressed that the NR C-V2X communication will be completed by the end of 2019 and will be commercialized by 2023 [16].

6.3.3 Hybrid V2X Technology

A common V2X platform is required that supports both the DSRC technology and C-V2X technology at the same time. If a single wireless communication technology is used, it cannot fulfill all the requirements of V2X communication suitable for intelligent and autonomous vehicles in ITS. In V2X, the DSRC is easy to deploy at a very low cost to support ad hoc communication as compared to cellular networks [17]. However, DSRC has certain limitations in highly dense traffic situations such as limited wireless range, short period of connectivity, and scalability issues [18]. The vehicles have a very high mobility; dynamic topology so the RSUs based on DSRC technology cannot provide wide coverage in urban city. In addition, during the early deployment of V2X, the RSUs may not be available in all parts of the country. On the other hand, cellular networks overcome these issues by providing wide wireless network coverage, high capacity, and high penetration in dense and urban city. However, the cellular networks also have few drawbacks before it can be applied in V2X communication. The cellular networks cannot handle high-frequency beacon messages from all the vehicles. There might be degradation in cellular networks service quality for safety messages if it has to support huge number of high mobility vehicles. In such type of situation, the cellular networks might not be suitable for sharing critical event messages due to latency issues.

In addition, if the vehicles on the roads are based on two different technologies like some vehicles are using DSRC technology and other vehicles are using C-V2X technology, then these vehicles based on two different types of communication technology cannot communicate with each other. As a result, the V2X communication potential cannot be obtained.

Hence, if we can combine both the technologies, it might be a worthy V2X platform. The hybrid V2X can meet several demanding communications requirements and can provide better ITS services in the future.

According to 5GAA [19], both the DSRC based on 802.11p and C-V2X have equal rights to operate in 5.9 GHz band considering the relevant technical regulatory conditions. In 3GPP Rel. 14, the coexistence of DSRC and cellular technologies in the same 5.9 GHz ITS band has been studied. The advantage of these studies is that it provides harmonized spectrum and low latency for V2X communication for intelligent and autonomous vehicles. In addition, if two technologies can combine, then it would provide synergy in V2X communication.

As 5G for C-V2X will be deployed soon, the radio access technology will be the combination of current and future licensed and non-licensed technologies such as LTE, 4G, 5G NR, Wi-Fi, and 802.11p. In case of autonomous and intelligent driving based on V2X, the multi-RAT technology may provide redundant connectivity or boost the capacity and throughput of V2N and V2I networks. The mm wave beamforming in cellular technology can be assisted by 802.11p technology, thus improving the efficiency of the networks. Hence, the hybrid V2X technology can provide advanced functionalities and orchestration for smooth traffic management and flow controlling.

6.3.4 C-V2X Applications and Requirements

The C-V2X application has been categorized into four types based on their requirements and use cases according to 5GAA. They are as follows:

(i) Safety: The safety use case such as intersection movement designed to reduce the accidents by warning the driver o f he approaching crash hazard.
(ii) Convenience: The convenience use case such as software updates like software Over the Air (OTA) updates for automotive management, vehicle health services as well as other telematics services for saving time.
(iii) Advanced Driving Assistance: In advanced driving assistance, the use cases such as real-time situational awareness and high definition maps alert driver regarding overtaking, icy road conditions in front, etc. It helps to improve the traffic flow by managing vehicle route, weather alerts, limiting speed, etc.
(iv) Vulnerable Road User (VRU). The VRU use case discovers vulnerable road users and warns the driver of VRUs about its location status. It maintains safe communication among vehicles and pedestrians, bicycles, bikes, and other users.

In this section, we discuss the service requirements of C-V2X to enhance the 3GPP systems in 5G scenarios. This comprises transport layer support for safety as well as non-safety C-V2X application [20]. The safety-related V2X applications are vehicles platooning, advanced driving, remote driving, and extended sensors. The non-safety-related V2X applications are infotainment, mobile hotspots, 3D HD dynamic digital map update, etc. Table 6.2 summarizes the requirements for autonomous vehicles.

6.4 Role of Edge Computing and SDN in V2X

The intelligent and autonomous vehicles will generate a massive amount of data such as sensor data, multimedia data, and other types of data that cannot be handled by traditional centralized servers. With the increase in the density of vehicles in urban

Table 6.2 V2X requirements for autonomous vehicles

Category		Communication mode	Latency (ms)	Throughput (Mbps)
Non-safety	Infotainment	V2I, V2N	500–1000	80
	Traffic safety	V2V, V2I, V2P	20–100	0.5–700
	Traffic efficiency	V2V, V2I, V2N	100–500	10–45
Safety	Advance driving	V2V, V2I	3–100	0.096–53
	Remote driving	V2N	5	25
	Vehicle platooning	V2V, V2I	10–20	0.012–65
	Extended Sensors	V2V, V2I, V2P	3–100	10–1000

areas and massive amount of data generated by them, the edge computing can satisfy the location awareness, mobility, and latency requirements [21]. In order to collect and process those massive amount of data instantaneously, edge cloud computing plays an important role. The edge clouds are located at the edge of the networks at the proximity of the vehicles and are geographically distributed. The edge cloud provides better performance, and services bring cloud-like facilities at the proximity of the vehicles. There are several concepts of edge cloud services among them is multi-access edge computing (MEC). Industry Specification Group (ISG) within ETSI introduced MEC. ISG is responsible for standardizing open environment and seamless integration of applications from vendors, service providers, etc., and focuses on uniting telecommunications and ITS sectors. MEC servers are deployed at several locations such as macrobase stations (eNB) or multi-radio access technology (RAT) [22] that performs cloud-like services at the edge of the network. The MEC provides applications and services such as storage, computing resources, edge connectivity, and network information based on cellular network. The computation, storage, and processing loads are handled in a distributed manner, instead of backhauling every bit of traffic into main datacenter [21]. This property is very useful for the V2X communication to deal with latency issues. It provides high bandwidth, ultra-low latency, location awareness, and access to the real-time information of HD maps. The MEC is being leveraged by 5G technology that provides additional benefits for multiple C-V2X use cases. The MEC infrastructure consists of virtualization layer that provides infrastructure-as-a-service (IaaS) capability and hardware resources. The MEC also supports smart vehicles with very high mobility with low latency. As the MEC is located at the edge of the network and in proximity to the vehicles, real-time data analysis can be achieved for processing high computational tasks such as HD maps, dynamic route discovery, and other infotainment messages at a very low latency. Recently, the 5GAA [23] and ETSI [24] have recognized the inclusion of MEC for cellular V2X communication.

Alongside the advancement in cellular 5G technology, software-defined network (SDN) plays an important role in V2X communication. SDN is an emergent technology that can be used in coordination with MEC. The centralized and organized paradigm of SDN offers flexibility, scalability, and programmability to V2X communication. The SDN is partitioned into two planes, i.e., control plane for network traffic

6.4 Role of Edge Computing and SDN in V2X

control and the data plane for the data forwarding function [25]. The most common protocol used for communication between the SDN control plane and data plane is OpenFlow [26]. The OpenFlow enhances the vehicles' resource management by allowing opportunity for new services and control functions. The basic SDN architecture consists of three layers, i.e., application layer, control layer, and network layer as shown in Fig. 6.5. The northbound interface interacts with SDN controller and application layer based on API and supports application developers to manage the network through the program. The southbound interface interacts with the SDN controller and network devices such as router, switches, and other network devices of the network layer. The SDN and other network virtualization function can be customized for special use cases providing efficient traffic performance.

The MEC computing combined with SDN plays a significant role in enhancing the performance of V2X communication [27]. By combining the SDN controller

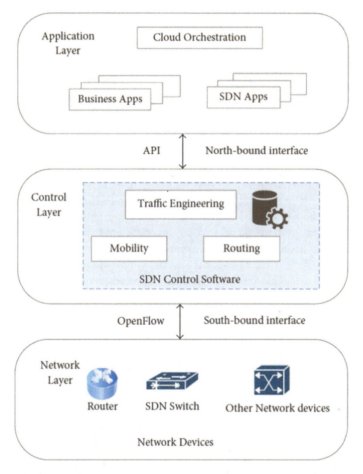

Fig. 6.5 SDN controller system

with MEC, the system can manage geographically distributed MEC devices and can manage heterogeneous networks by providing flexibility, programmability, and overall information about the network. Due to the open and flexible setup interface of SDN, it offers intelligence, which can be used in V2X communication. In addition, the decentralized nature of MEC decreases delay and enhances resource utilization in V2X. If we consider example of hybrid V2X communication, the SDN controller can divide the task between the eNB and RSUs. The SDN wireless nodes can be the vehicles that use cellular technology for control channel and DSRC for the data channel. In case, the vehicles need to send the emergency message to other vehicles in the network, and it will broadcast the message. The edge servers at the MEC will analyze and process the information. Based on the received emergency message's event type, contents, timeliness, and priority, the servers will transmit the emergency messages to other MEC devices that are geographically close to the accident location. The MEC server can send the gathered information to the cloud for big data analysis, and the results can be provided to the third parties such as insurance company, police departments, or hospitals for further processing. Hence, using dedicated SDN virtualization, and orchestrating MEC distributed networks for V2X communication, the safety and critical applications can be prioritized over different network traffic.

6.5 Connected Vehicle Cyber Security

The term cybersecurity refers to the technology, processes, or practices designed to protect networks, computers, programs, and information against cyberattacks. The cybersecurity protects the system or networks from malicious cyberattacks that interrupt the normal communication in the network or thwart the functioning of the system or steal the sensitive information. This section discusses about the cybersecurity of the intelligent and autonomous vehicles against different types of attack vulnerabilities, hacking, associated risks, their preventions, and solutions. We will discuss the different types of security and privacy issues and security requirements in connected vehicles. Then, we discuss the trust management issues, homomorphic encryption, and blockchain as a security in V2X communication. This section especially focuses on security and functional safety related to V2X communication.

6.5.1 WAVE Communication Cybersecurity

In intelligent transportation system, the WAVE is used as a communication mode by vehicles to run in the DSRC band. In the USA, the SAE J2945/1 and J2945/2 standards were defined for safety requirements of on-board systems for V2V communication and DSRC performance requirements for V2V safety awareness, respectively. The J2945/1 specifies the on-board V2V communications security, functional, and performance requirements for light vehicles. The J2945/2 specifies DSRC

6.5 Connected Vehicle Cyber Security

safety requirements including detailed system engineering documentation as well as interoperability for V2V communication safety awareness.

The safety-related application in V2X communication is very time sensitive, so the bandwidth overhead and processing time should be kept minimum while maintaining security and privacy. The security requirements defined in ETSI TR TR102893 and SAE J2945/1 meet the application security and management messages described by the IEEE 1609.2 standard. The IEEE 1609.2 standard outlines the security service provided by MAC layer for applications performed on the WAVE network stack. The WAVE security services consist of WAVE internal security services and WAVE higher layer security services [28] as shown in Fig. 6.6. The WAVE Internal Security Services are secure data service (SDS) that transforms unsecured protocol data units (PDUs) to secured protocol data units (SPDUs) to be transferred between nodes. The security management manages the information related to the certificates. The WAVE higher layer security services are (i) certificate revocation list (CRL) verification entity (CRLVE) that validates incoming CRLs and permits the associated revocation data to the SSME for storage and (ii) peer-to-peer certificate distribution (P2PCD) entity (P2PCDE) that allows peer-to-peer certificate distribution.

In WAVE, the encryption technique guarantees the essential security requirements such as authenticity, integrity, confidentiality, and anonymity. In IEEE 1609.2 standard, the security functions such as confidentiality, integrity, availability, authentication, and authorization are delivered through security services.

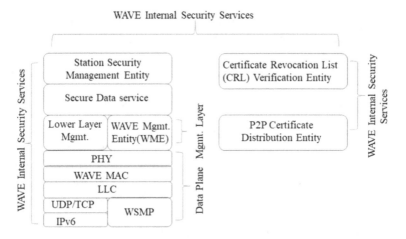

Fig. 6.6 WAVE protocol stack showing WAVE security services

6.5.2 Security and Privacy in V2X Communication

Although V2X communications offer safety and environmental benefits, there are security and privacy concerns. In V2X communication, privacy and authentication are of utmost importance that provide full security to the vehicular nodes. The system verifies the identity of the vehicle user while providing privacy to the user's private data, i.e., his private and confidential information is not be revealed. However, the user should be traceable by the authority in case of legal issues like accidents, hit and run, etc.

- *Authentication*: In V2X communication, if the message or data arrives from unauthorized or unreliable source, it might contain misleading or harmful information, and it should be handled carefully. Therefore, only authorized nodes can transmit the message, and only the receiving node can process the data whose data origin has been verified. The V2X nodes should authenticate the source of the received message; check the confidentiality and integrity of the message to prevent from replay attacks. Authentication process is one of the important security requirements of V2X communication to maintain trust among the nodes. There should be strong authentication scheme in V2X as one weak vehicular node may compromise the security of the entire V2X system.
- *Integrity*: In V2X, the received message by the vehicular nodes should preserve message integrity for the legitimate messages. Some malicious vehicular nodes try to inject fake messages by using message modification attack that modifies whole or some part of the legitimate message. This results in taking the attacker intended actions by nearby vehicular nodes.
- *Non-repudiation*: It is one of the important requirements for security in V2X communication. It guarantees that the message source vehicle cannot deny any messages transmitted by the sender or by the receiver.
- *Anonymous authentication*: The anonymous authentication protects unauthorized vehicular nodes from gaining access to the system and prevents from unauthorized attackers. We need to consider the privacy while designing anonymous authentication for the vehicular nodes.
- *Anonymity*: Anonymity of the individual vehicular nodes should be maintained while communicating with other nodes or infrastructure so that his real identity is not disclosed as well as privacy is maintained. To preserve the identity privacy, the nodes should send the data anonymously so that his identity cannot be traced. However, in case of the node behaves maliciously, its identity can be detected.
- *Traceability*: Since the V2X communication is accomplished anonymously, it maintains the anonymity. However, if malicious user tries to misbehave malevolently, then that user needs to be detected or traced by the system. The authority needs to track and recognize the malicious user's actual identity in order to cancel or revoke the membership from the system.
- *Unlinkability*: The real identity of the vehicular nodes should not be linked with his location or other parameters. The vehicle should use a temporary identity known as pseudo id (PID) that is heterogeneous to the real identity of the vehicle. The

temporary identity of the vehicle should change frequently overtime to achieve unlinkability. Even if the malicious entities attempt to obtain his PID, they could not relate them to the real identity. This unlinkability protects the vehicle from tracking.
- *Confidentiality*: It is one of the important requirements of privacy for entity trust in ITS system. Confidentiality provides identity and location privacy of the vehicular nodes. V2X communication should provide message confidentiality, which guarantees that the messages are read by the authorized vehicles only. It protects from identity theft and location tracking.

Table 6.3 shows a list of threat, attacks, and their tentative solutions in intelligent and autonomous vehicles. We will briefly discuss the different types of attacks in three components of IAV, and details can be found in [29].

6.6 Trust Management in V2X Communication

In V2X, trust enhances security, and it is one of the essential components in creating a trusted V2X communication. The trust between vehicular nodes as well as trust on the received message plays an important role to maintain security in the vehicular communication. The trust management system in V2X communication enables the vehicular nodes to identify malicious vehicles as well as detect fake messages sent by malicious vehicles. It can also impose punishment on the malicious nodes by giving low trust score so that they behave honestly and share legitimate information between the vehicles in the future. There are several trust management schemes (TMS) and trust models in vehicular networks. The trust management schemes evaluate the trust values of the neighbor vehicles to prevent them from interacting with the malicious vehicles. The TMS are divided into four categories based on the use of infrastructure and cryptographic measures such as public key infrastructure (PKI) as described in Chap. 3. The trust management schemes based on infrastructure use RSU or central authority for trust evaluation but without depending upon PKI. Similarly, the trust management system based on infrastructure and PKI can effectively identify attacker with some degree of accuracy. There are researchers who work on solving the issues in message trustworthiness based on PKI but without depending upon the infrastructures. On the other hand, TMS that work in a distributed environment and infrastructure less environment without using PKI help to overcome the limitation of the other trust management system.

The existing trust management schemes are based on particular application area applying different trust-based models to enhance V2X communication. The trust-based models are classified into three categories. They are entity-based, data-centric-based, and hybrid trust models as shown in Fig. 6.7.

In entity-based trust model, it deals with the trust of each vehicle and verifies the trustworthiness considering the opinions of peer vehicles. The trust management between two nodes depends on two types of observations, i.e., direct observation and

Table 6.3 Threat, attacks, and solutions in intelligent and autonomous vehicles

Components			Threats	Attacks	Solutions
Vehicular System	Vehicle		Unauthorized routing table control	Jamming attacks	Frequency hopping, multiple radio transceivers
			Exposure of sensitive data	Sensor impersonation	SPECS [30]
			Network flooding with false data	Bogus information	ECDSA [31]
			Illegitimate software updates	Remote access attack	FOTA [32]
			Damaging sensors	Physical attack	Access control
	Vehicle/driver		Privacy leakage of personal data	Malware integration	Updating antivirus, sandbox approach [33]
			Privacy leakage of sensitive data	Social engineering attack	Encrypted and strong password for message communication
	Driver		Users identity exposure	User privacy exposure	Holistic approach for data transmission [34]
	Commn. network (wired)		Sensitive data exposure	Eavesdropping between central entity and RSU	Encrypted messages
			Message exchange during transmission	MITM attacks between central entity and RSU	Strong cryptographic techniques
			Messages disposal	Wormhole attacks	Packet leash [35], HEAP [36]
Information	Exchange message		Exposure of sensitive data and user's personal information	Eavesdropping	Strong encrypted message for user's communication
			Blocks vehicles from receiving critical message and access network services	Jamming attacks	Assign secure IPs to vehicles while exchanging message [37], DJAVAN [38]

(continued)

6.6 Trust Management in V2X Communication

Table 6.3 (continued)

Components		Threats	Attacks	Solutions
		Messages modification	Impersonation attacks	Batch verification scheme based on Identity [39]
		Message modification with false and compromised messages	MITM attacks	Strong cryptographic techniques
		Message manipulation and dropping	Spoofing attacks	Multi-antenna receiver against movements, secure location verification [40]
Infrastructure	RSU, central entity	Leakage of information on back-end wired channel	Sybil attacks	Autonomous neighbor nodes position verification [41] sybil attack detection [18], RobSAD [42]
		Network flooding with compromised messages	False message between central entity and RSU	ECDSA

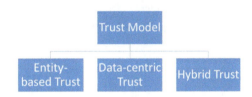

Fig. 6.7 Different types of trust-based models

indirect observations. In direct observation, the evaluating node directly observes the peer nodes behaviors. In indirect observation, the observations are collected by using direct observations with neighbor nodes in vehicular networks at the cost of overhead due to observation collection. The authors in [43] proposed a fuzzy method for the verification of the trustworthiness of the nodes by using feedback from its peers. Nevertheless, the trustworthiness of a message may not always agree with the trustworthiness of the node itself. In addition, the trust based on the history of interactions is not feasible for vehicular networks due to its dynamic nature. Also, they do not distinguish between the trust value of a node and that of the message and considered as equivalent, but trustworthiness of nodes and messages is not same. In case of data-centric trust model, the trustworthiness of the received event messages from the neighbor vehicles is calculated instead of trust of the entities or the vehicle itself. In hybrid trust model, the entity-based and data-centric trust models are combined to estimate the trustworthiness of a node as well as trustworthiness of message. The authors in [44] proposed a hybrid beacon-based trust management system (BTM). It is based on entity trust from beacon messages and computes data trust from validating the credibility of event messages and beacon messages. They have used the direct event-based trust and indirect event-based trust to compute the data-centric trust. The receiver vehicle is able to evaluate the trustworthiness of the sender vehicle by analyzing the beacon message as well as event messages from a vehicle. They implement the cosine similarity to calculate entity-based trust from beacon messages.

6.7 Homomorphic Encryption in VANET

The homomorphic encryption (HE) is a type of encryption that permits sender to encrypt his information using cryptographic keys. The HE allows the third party to achieve certain type of mathematical operations on the encrypted information without decrypting it, at the same time maintaining the privacy of the sender's encrypted information. In 2009, Craig Gentry [45], elaborated the first possible construction for a fully homomorphic encryption using lattice-based cryptography. It allows both addition and multiplication operations on ciphertexts, from which it is capable of constructing circuits for accomplishing arbitrary calculation. In HE, the mathematical operation on the plaintext during encryption is equivalent to another operation performed on the ciphertext. There are three types of homomorphic encryption based on the mathematical operations on the encrypted data. The three types of homomorphic encryption schemes are shown in Fig. 6.8, and the description is given below.

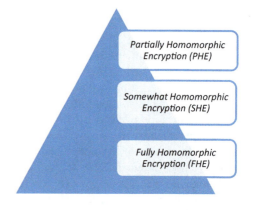

Fig. 6.8 Types of homomorphic encryption schemes

6.7 Homomorphic Encryption in VANET

(a) *Partially Homomorphic Encryption (PHE)*: Partially homomorphic encryption comprises schemes that support the evaluation of only one type of mathematical operation on the encrypted message, i.e., it permits either addition or multiplication operation, with unrestricted number of times. The PHE can be used only for certain types of applications.

(b) *Somewhat Homomorphic Encryption (SHE)*: Somewhat homomorphic encryption scheme can evaluate two types or some simple mathematical operations on the encrypted messages, i.e., it permits both addition and multiplication operation, but with restricted number of times. The drawback of SHE scheme is that it permits only limited number of homomorphic computation due to increase in size of the ciphertext along with each homomorphic computation.

(c) *Fully Homomorphic Encryption (FHE)*: Fully homomorphic encryption scheme supports the evaluation of huge number of different types of operations on the encrypted message with unrestricted number of times. The FHE scheme is composed of partially homomorphic encryption and somewhat homomorphic encryption. However, FHE is not suitable for real-world situation due to its great complexity and calculation time for handling arbitrary number of homomorphic operations for unrestricted number of times. Many researchers are planning privacy-preserving, secure, and practical FHE that takes low computation time suitable for real-world situation.

The homomorphic encryption can be used in different ways for providing security and privacy in vehicular networks. The HE can be used as a security mechanism to preserve the location privacy and movement of the vehicles. The homomorphic encryption can be used to encrypt the location and direction information before sharing the location information to the third party. Then, HE scheme is used to evaluate and verify the location and direction information. The comparison results will not reveal any information to the third party. Hence, HE provides security against the malicious vehicles and provides security and privacy to the vehicles. Similarly, the vehicles can use HE to encrypt their route before reporting to other vehicles and RSUs. At the RSU side, the RSUs can verify and process multiple route reporting packets received from multiple vehicles by using batch verification technique in very short time. After verification of the collected information, the RSUs send the route reports to the Traffic Management Center (TMC). In this way, the third party cannot know the content of the encrypted route information, hence preserving privacy of the vehicles. Similarly, the authors in [46] proposed a novel randomized authentication (RAU) protocol based on Paillier's public key for maintaining privacy. The RAU protocol is appropriate for mobile ad hoc as well as vehicular network settings. Their scheme is based on zero-knowledge proof to achieve anonymous authentication. Each user self-generates a large number of authenticated identities to preserve privacy. The authentication servers are used to trace and identify the attacker nodes if there are any disputes or malicious activities. It is suitable for small IoT devices.

6.8 Blockchain in V2X Communication

Within the last decade, blockchain has gained a lot of attention and recognition from academia, IoT industry, automotive industry, financial technology, health care, real state, etc. The well-known Bitcoin cryptocurrency [47] gives rise to a decentralized and distributed computing technology identified as blockchain. Blockchain can be a good candidate as a security solution for smart cities. The main reason behind the popularity of blockchain is the growth in the market capitalization of cryptocurrencies that in turn generates various blockchain applications. Blockchain has a strong potential to create big impact on the distributed and decentralized ledger system. The current centralized database system suffers from scalability and single point of failure as the number of nodes and transaction increases. There are different variants of blockchain that can handle the shortcomings of traditional database system. One of the big advantages of blockchain is that it can work in a distributed and decentralized trustless environment. The nodes need not trust each other in the first place to carry out their transactions while maintaining privacy that can overcome the trust-related issues. It provides security and privacy in peer-to-peer networks.

There are several useful features of blockchain. Some of the features of blockchain are they are distributed, decentralized, anonymous, transparent transaction, immutable, traceable, and non-repudiation. In ITS, along with the advanced features and evolution in vehicles, they are becoming more intelligent and autonomous. The autonomous vehicles send and receive information with different sensors as well as neighbor vehicles and RSUs. The exchange of information between adjacent vehicles is susceptible to various types of attacks. The blockchain can be a solution to secure the communication between trustless vehicular nodes in V2X communication [48]. The consensus mechanism in blockchain acts as a security mechanism among untrustworthy vehicles in the trustless vehicular environment. Based on the access mechanism, the blockchain can be of two types, and they are permissioned and permissionless blockchain. The difference between permissioned and permissionless blockchain is given in Table 6.4.

The vehicular network still suffers from security and privacy issues. There are many security mechanisms that exist for securing VANET but none of them can provide adequate security and privacy. Some security techniques are successful in securing the vehicular communication from external attacks; however, they cannot prevent from insider attacks. There are some attacks that analyze the broadcasted messages between the vehicles and modify the message that cause severe effects such as de-route the vehicle from its original destination. This causes serious accidents as well as takes the life of the driver. The blockchain can be used to store the traffic event logs as transactions. The blockchain can also be used to record vehicle node trustworthiness and message trustworthiness that acts as a ground truth for forensics whenever there is a serious accident.

The blockchain can be used as a distributed solution for smart vehicle security and privacy. The authors in [49] used the lightweight scalable blockchain (LSB)

6.8 Blockchain in V2X Communication

Table 6.4 Difference between permissioned and permissionless blockchain

Permission less Blockchain	Permissioned Blockchain
Open access	Authorized access: Presence of access control layer in the BC nodes
It focuses on censorship-resistant and anonymous transactions	It focuses on financial regulations
Nodes are anonymous and fully decentralized validators	Nodes are preselected and trusted validators
Permissionless transaction verification	Network participants have ability to restrict who can create smart contract and/or transact on BC network
Almost impossible to reverse transactions once written on-chain	Certain network nodes have control to undo or edit transactions
Can take part in consensus mechanism	Restriction to take part in consensus mechanism
Highly secure as miners are provided with incentives and use consensus mechanism such as PoW	Less secure; collusion of authorized nodes and brute force attack of 2/3rd of private keys for validator set
Substantial computation power is required to maintain large distributed network, no privacy for transaction	Scalable computation power and transaction throughput
Source code is open and anyone can propose upgrades	Network nodes may be contractually bound to implement network upgrades
Existing cryptocurrencies, e.g., Bitcoin	Enterprise-level systems, business, financial, health care, etc., e.g., hyperledger fabric

blockchain concept to reduce blockchain overhead and provide decentralized security and privacy by using overlay networks for smart vehicles. They used symmetric encryption scheme for all the transactions. The information is exchanged in a trustworthy manner so that security of the vehicles can be preserved. The privacy of the vehicles is maintained by locally recording location data of the vehicles. The blockchain can be used for forensic applications as well. The authors in [50] used lightweight permissioned blockchain in connected vehicles. The blockchain records traffic information like as road accidents, vehicle repair history, insurance, and vehicle diagnostic reports, etc. The authors integrate vehicular public key management (VPKI) scheme in the private blockchain for membership registration in a distributed ledger that offers security and privacy to the vehicles. This scheme consumes very less memory and low processing overhead, but the distributed ledger stores hash values only and does not guarantee information accuracy and information availability.

6.9 Safety Standards for IAV

Safety and security of the passengers are the most important aspect of intelligent and autonomous vehicles because it involves the human life. The drivers and passengers place their safety in the hands of the autonomous vehicles by trusting their security system. So, the IAV requires a new level of safety standards on which the drivers can trust. The safety standards should guarantee high level of security, integrity, and efficient operation of the autonomous vehicles. The IAV security objectives can be derived based on the relevant autonomous vehicle safety standards because the safety issues directly lead to the criticality of the cybersecurity. Some of the existing safety standards for autonomous vehicles are briefly discussed below:

(a) International Organization for Standardization—ISO 26262: The ISO 26262 provides a safety life cycle during the course of automotive product development phases such as development, management, production, operation, service, and decommissioning [51]. The ISO 26262 is an adaptation of the functional safety standard International Elecrotechnical Commission, IEC 61508 that was originally developed for automotive electronic and electrical safety-related systems [52]. In addition, the ISO 26262 defines the Automotive Safety Integrity Level (ASIL), which is functional safety for road vehicle safety standards. It includes different types of classification of the safety life cycle. The ASIL is responsible for the hazard analysis and risk assessment process that identify all the different types of hazardous events and safety goals. The severity (S) is classified based on the hazardous events and its expected severity of injuries. It is classified from $S0$ to $S3$, where $S0$ represents no injuries while $S3$ is life threatening injuries. The likelihood of injuries due to exposure (E) to the hazardous events is classified from $E0$ to $E4$ where $E0$ represents unlikely to happen injuries and $E4$ represents high probability of injury. Similarly, the controllability (C) is the relative probability the driver can control the situation to stop the injury. It is also classified from $C0$ to $C3$, where $C0$ is the general control while $C3$ is uncontrollable. In general, the ASIL is represented as below

$$\text{ASIL} = \text{Severity} \times \text{Exposure} \times \text{Controllability}$$

The higher the level of ASIL, the situation is more threatening and might cause severe damage. In terms of intelligent and autonomous vehicle, the controllability level reaches three, i.e., highly uncontrollable. There are several other techniques to assess the ASIL level such as fault tree analysis (FTA), hazard analysis and risk assessment (HARA), and failure mode and effects analysis (FMEA) [53].

(b) SAE J3061: The severity, control, and exposure explanations provided by the ASIL hazard classification are only for information purpose but it is not enough for practical safety standards. The Society for Automotive Safety Engineers (SAE) introduced SAE J3061 that delivers specific guidance for assessing the abovementioned hazard classification like severity, control, and exposure for a particular hazard that are suitable for various automakers and vehicle components suppliers. It provides a high-level guidance on vehicle cybersecurity and is adaptable in vehicle industry like cyber physical vehicular systems. The SAE J3061 considers the ISO 26262 and creates the awareness and general terminology in autonomous vehicle supply chain. It specifies the distinction between the system cybersecurity and system safety as the cybersecurity is related to the persistent attack while the system safety is related to the fault or accident of the vehicles. It acts as a foundation for future standards in the area of vehicle cybersecurity.

6.10 Summary

This chapter provides in-depth knowledge on types of inter-vehicle communication. It gives a clear definition of DSRC and cellular vehicular networks and their evolution and adoption in the autonomous vehicular environment. More specifically, this chapter introduces unique features of the cellular V2X based on 5G technology such as data control software-defined network (SDN), scalable network architecture and topology, edge cloud computing like cloud/fog computing and processing, and application-oriented design as part of smart vehicles. In addition, it discusses about the cybersecurity of the intelligent and autonomous vehicles against different types of attack vulnerabilities, hacking, associated risks, their preventions, and solutions along with different types of security and privacy issues and security requirements in connected vehicles. It also includes trust management issues, homomorphic encryption, and blockchain as a security in V2X communication. This section especially focuses on security and functional safety related to V2X communication.

References

1. DoT, "Connected Vehicles." [Online]. Available: https://www.its.dot.gov/cv_basics/index.htm
2. K. Sjoberg, P. Andres, T. Buburuzan, A. Brakemeier, Cooperative intelligent transport systems in Europe: current deployment status and outlook. IEEE Veh. Technol. Mag. **12**(2), 89–97 (2017)
3. 3 GPP, Technical specification group services and system aspects. Study on architecture enhancements for EPS and 5G system to support advanced V2X services, 3GPP TR 23.786 v0.8.0., (2018)
4. Y.L. Morgan, Notes on DSRC & WAVE standards suite: Its architecture, design, and characteristics. IEEE Commun. Surv. Tutorials **12**(4), 504–518 (2010)
5. ETSI, Intelligent Transport Systems (ITS), Access layer specification for Intelligent Transport Systems operating in the 5 GHz frequency band, (France, 2013)
6. V.D. Khairnar, K. Kotecha, Performance of vehicle-to-vehicle communication using IEEE 802.11p in vehicular ad-hoc network environment. Int. J. Netw. Secur. Its Appl. **5**(2), 1–28 (2013)
7. Technical specification group services and system aspects, Security aspect for LTE support of Vehicle-to-Everything (V2X) services, 3GPP TS 33.185 V14.1.0., (2017)

8. 3 GPP, Technical specification group radio access network. Evolved universal terrestrial radio access (E-UTRA), Physical layer procedures, Rel. 15, TR 36.213 v15.3.0, (2018)
9. 3 GPP, Technical specification group services and system aspects, Release 14 Description, Summary of Rel–14 Work Items, (2018)
10. A. Haider, S.-H. Hwang, Adaptive transmit power control algorithm for sensing-based semi-persistent scheduling in C-V2X Mode 4 communication, Electronics, **8**(8), (2019)
11. 3 GPP, LTE: service requirements for V2X services, 3GPP TS 22.185 version 14.3.0 Release 14, (2017)
12. 5G Automotive association, The case for cellular V2X for safety and cooperative driving, *5GAA Whitepaper*, pp. 1–8, (2016)
13. Autotalks, DSRC vs. C-V2X for safety applications, [Online]. Available: DSRC vs. C-V2X for Safety Applications. [Accessed: 01-Nov-2019], (2019)
14. ITU-R M.2410–0, Minimum requirements related to technical performance for IMT-2020 radio interface(s)
15. X. Wang et al., Millimeter wave communication: a comprehensive survey. IEEE Commun. Surv. Tutorials **20**(3), 1616–1653 (2018)
16. A. Frost, 5GAA says C-V2X is ready to roll out globally this year, https://www.traffictechnologytoday.com
17. R. Shrestha, R. Bajracharya, S. Djuraev, S.Y. Nam, K.-S. Sohn, Performance evaluation of heterogeneous VANET based on simulation. Far East J. Electron. Commun. **2**(1), 155–162 (2016)
18. R. Shrestha, Sybil attack detection in vehicular network based on received signal strength, 1, (2014)
19. 5GAA, Coexistence of C-V2X and ITS-G5 at 5.9 GHz, https://5gaa.org/, 2018. [Online]. Available: https://5gaa.org/wp-content/uploads/2018/10/Position-Paper-ITG5.pdf. [Accessed: 29-Oct-2019]
20. 3GPP, Service requirements for enhanced V2X scenarios, 2018
21. R. Shrestha, R. Bajracharya, S.Y. Nam, Challenges of future VANET and cloud—based approaches, Wirel. Commun. Mob. Comput., 2018, no. Article ID 5603518, 1–15 (2018)
22. H.Y. Chao, M. Patel, D. Sabella, N. Sprecher, V. Young, Mobile edge computing a key technology towards 5G, ETSI (European Telecommun. Stand. Institute), pp. 1–16 (2015)
23. D. Sabella et al., Toward fully connected vehicles: edge computing for advanced automotive communications, 2017
24. ETSI, Multi-access Edge Computing (MEC): study on MEC support for V2X use cases, 2018
25. Open Network Foundation, *White paper: software defined networking: the new norm for networks* (Palo Alto, Calif, USA, 2012)
26. H. Kim, N. Feamster, Improving network management with software defined networking. IEEE Commun. Mag. **51**(2), 114–119 (2013)
27. X. Duan, Y. Liu, X. Wang, SDN enabled 5G-VANET: adaptive vehicle clustering and beamformed transmission for aggregated traffic. IEEE Commun. Mag. **55**(7), 120–127 (2017)
28. IEEE SA, IEEE 1609.2–2016—IEEE Standard for wireless access in vehicular environments–security services for applications and management messages, 2016
29. F. Ahmad, A. Adnane, V.N.L. Franqueira, A systematic approach for cyber security in vehicular networks. J. Comput. Commun. **04**(16), 38–62 (2016)
30. T.W. Chim, S.M. Yiu, L.C.K. Hui, Z.L. Jiang, V.O.K. Li, SPECS: secure and privacy enhancing communications schemes for VANETs. Ad Hoc Netw. **9**(2), 160–175 (2011)
31. S. Manvi, M. Kakkasageri, D. Adiga, Message authentication in vehicular ad hoc networks: ECDSA based approach, Int. Conf. Future Comput. Commun., pp. 16–20 (2009)
32. D.K. Nilsson, U.E. Larson, A defense-in-depth approach to securing the wireless vehicle infrastructure. J. Networks **4**, 552–564 (2009)
33. N. Nikaein, S.K. Datta, I. Marecar, C. Bonnet, Application distribution model and related security attacks in VANET, Int. Conf. Graph. Image Process., 2012
34. K.S. Tamil Selvan, R. Rajendiran, A holistic protocol for secure data transmission in VANET, Int. J. Adv. Res. Comput. Commun. Eng., **2**, 4840–4849 (2013)

35. Y.C. Hu, A. Perrig, D. Johnson, Packet leashes: a defense against wormhole attacks in wireless networks, in *22nd Annual Joint Conference of the IEEE Computer and Communications*, pp. 1976–1986 (2003)
36. S.M. Safi, M. Movaghar, A. Mohammadizadeh, Novel approach for avoiding wormhole attack in VANET, in *1st Asian Himalayas International Conference on Internet*, pp. 54–59 (2009)
37. K. Verma, H. Hasbullah, A. Kumar, Prevention of DoS attacks in VANET. Wirel. Pers. Commun. **73**, 95–126 (2013)
38. L. Mokdad, J. Ben-Othman, A. Tuan Nguyen, DJAVAN: detecting jamming attacks in vehicle ad hoc networks, Perform. Eval., 87(C), pp. 47–59 (2015)
39. C. Zhang, R. Lu, X. Lin, P.H. Ho, X. Shen, An efficient identity-based batch verification scheme for vehicular sensor networks, in *The 27th IEEE Conference on Computer Communications*, pp. 816–824 (2008)
40. T.E. Montgomery, P.Y. Humphreys, B.M. Ledvina, Receiver-autonomous spoofing detection: experimental results of a multi-antenna receiver defense against a portable civil GPS spoofer, in *Proceedings of the ION International Technical Meeting*, pp. 124–130 (2009)
41. T. Leinmuller, C. Maihofer, E. Schoch, F. Kargl, Improved security in geographic ad hoc routing through autonomous position verification, in *Proceedings of the 3rd International Workshop on Vehicular Ad Hoc Networks*, pp. 57–66 (2006)
42. C. Chen, X. Wang, W. Han, B. Zang, A robust detection of the sybil attack in urban VANETs, in *29th IEEE International Conference on Distributed Computing Systems Workshops*, pp. 270–276 (2009)
43. F. Gómez Mármol, G. Martínez Pérez, TRIP, a trust and reputation infrastructure-based proposal for vehicular ad hoc networks, J. Netw. Comput. Appl., **35**(3), 934–941 (2012)
44. Y.M. Chen, Y.C. Wei, A beacon-based trust management system for enhancing user centric location privacy in VANETs. J. Commun. Networks **15**(2), 153–163 (2013)
45. C. Gentry, A fully homomorphic encryption scheme, Stanford University, 2009
46. W. Jiang, D. Lin, F. Li, E. Bertino, Randomized and efficient authentication in mobile environments, Cyber Center Publication, pp. 1–15, 2014
47. S. Nakamoto, Bitcoin: a peer-to-peer electronic cash system, www.bitcoin.org, 2008
48. S. Kim, *Blockchain for a Trust Network Among Intelligent Vehicles*, 1st edn. vol. 111. (Elsevier Inc., 2018)
49. A. Dorri, M. Steger, S. Kanhere, R. Jurdak, BlockChain: a distributed solution to automotive security and privacy, IEEE Commun. Mag. Mag., **55**(12), 119–125 (2009)
50. M. Cebe, E. Erdin, K. Akkaya, H. Aksu, S. Uluagac, Block4Forensic: an integrated lightweight blockchain framework for forensics applications of connected vehicles. IEEE Commun. Mag. **56**(10), 50–57 (2018)
51. SAE International, Taxonomy and definitions for terms related to driving automation systems for on-road motor vehicles J3016_201806, *SAE International*. [Online]. Available: https://www.sae.org/standards/content/j3016_201806/. [Accessed: 20-Mar-2020]
52. IEC, Functional safety and IEC 61508, International Electrotechnical Commission, 2020. [Online]. Available: http://www.iec.ch/functionalsafety/. [Accessed: 16-Jan-2020]
53. A. Chattopadhyay, K.-Y. Lam, Autonomous vehicle: security by design, 2018

Chapter 7
Internet of Vehicles, Vehicular Social Networks, and Cybersecurity

7.1 Overview

In this chapter, we will discuss the emerging mobility ecosystem and its potential growth that enhance the vehicle movement from source to the destination. This helps to save the passenger's time and provides comfort during traveling. The main objective of the future of mobility along with the autonomous transportation and shared mobility is to overcome the predictable accidents on the road. The vehicles have evolved from mechanical transportation means to the smart vehicles with varieties of communication and sensing capabilities. With the development of Internet of Things (IoT), the conventional vehicles have developed over time and turned to Internet of Vehicles (IoV) with advanced connectivity with Internet, infrastructures, context awareness, sensing capabilities, service provisioning, etc. The IoV has developed over time from the conventional vehicular networks that connect the smart vehicles to the smart city. The IoV is a multifaceted vehicular network, where the vehicles consist of various sensors installed that gather data from other vehicles and road infrastructures. IoV is a special case of IoT, which is developed specially for automotive vehicles. Several applications of IoV are indispensable for vehicles, drivers, pedestrian, smart city infrastructure, etc. The IoV has been popular among the automotive industry research, and they are attracted to the academic sector as well. In IoV, a large number of vehicles interconnect with each other and they communicate through heterogeneous wireless communications. We discuss seven-layered architecture of IoV in detail. The seven layers of IoV are interface layer, data acquisition layer, data preprocessing layer, communication layer, management and control layer, processing layer, and the security layer. In IoV, there exists congestion and scalability issue due to the increase in the number of vehicles and are not able to preprocess huge amount of collected data from various sensors and neighbor vehicles. The IoV requires serious attention, and the security in the IoV needs to be studied carefully to prevent from serious cyber-attacks. The existing IoV architecture lacks security requirements such as authentication, authorization, and trust-related issues. We discuss several types of attacks in IoV and challenging issues in IoV. Then, we mentioned three types of IoV applications in intelligent transportation system (ITS), business-related applications, and smart city applications.

It is not easy to implement IoV in smart city due to several issues such as dynamic mobility, dynamic vehicle density, changing network topology, heterogeneous communication, and delay. The machine learning stands as a suitable solution to overcome these issues. Machine learning can process and utilize the information generated, collected, and stored by a variety of sensors. They exploit the information generated and stored in the network like vehicle's behavior patterns, locations, mobility, and network topologies to learn and extract the features. We discuss several types of machine learning techniques in IoV such as supervised, unsupervised, semi-supervised, reinforcement, and deep learning. We provide few detailed applications of machine learning techniques in IoV.

7.2 Internet of Vehicles (IoV)

Today Internet plays an important part in our daily activities similar to electricity few decades before. Without Internet, we cannot carry out our daily official activities whether it is a big or small organization. The Internet has reached almost all the parts of the globe. Many people can access the Internet and get huge benefit from it. Recently, various electronic devices that are known as "Things" can connect to the Internet and are able to communicate with other various devices providing different services. The network of such resource constrained heterogeneous things connected to the Internet and exchange information is known as "Internet of Things" in short IoT as shown in Fig. 7.1 [1]. The IoT devices have very low computing power, low memory, and limited battery life. The IoT has enormous potential to deliver diverse types of services across different fields from industry, business, intelligent transportation system (ITS), social media, health care, and smart cities. The IoT will have a deep impact on Internet of Vehicles, Internet of drones, etc. Since every sector has their own smart things connected over the Internet, we can call it as Internet of Everything (IoE) [1, 2].

With the advancement of IoT, a new paradigm in the field of ITS has emerged called Internet of Vehicles (IoV). The IoV has developed over time from the conventional vehicle ad-hoc network (VANET), infrastructures as well as other road transport equipment. The IoV is a complex vehicular network connected with the Internet with various types of sensors installed that gathers data from other vehicles and road infrastructures. Different from IoT system, the IoV has its own specific characteristics. The characteristics of IoV are as follows:

(a) *Complex communication*: The communication in IoV is based on different types of internal and external sensors installed in the vehicles such as LiDAR, radar, GPS, cameras, parking, brake, fuel, and temperature sensors. In addition, the vehicles use beacon messages and safety messages to communicate with other vehicles and network devices. The communication complexity increases with

7.2 Internet of Vehicles (IoV)

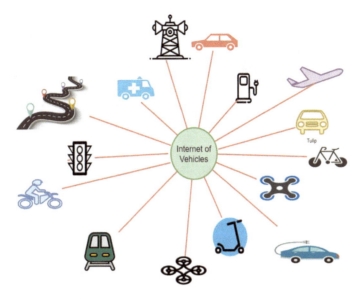

Fig. 7.1 Internet of vehicles

the change in density of vehicles in IoV network from urban scenario to highway scenarios. In highway scenarios, there will be fewer vehicles and they move a very high speed. However, in urban scenario, the vehicles will move at low speed but there will be interferences from neighbor vehicles. There is tradeoff between the density and mobility of the vehicles. The IoV can handle communication complexity in vehicular networks.

(b) *Dynamic topology*: In IoV, various types of sensors and heterogeneous components communicate with each other. The vehicles move at a very high speed on the road that changes the network topology quickly. The IoV can handle the dynamic topology of the vehicular networks. In addition, the density of the vehicles varies according to geographical location, city, and highway. Therefore, the IoV has non-uniform density as well that leads to dynamic topology.

(c) *High scalability*: The density of the vehicles is increasing rapidly in suburban location. In smart city, the vehicles are becoming intelligent and autonomous with various types of sensors installed on them. The IoV can handle the ever-growing number of vehicles and extend the network in large-scale environment providing high scalability.

(d) *Localized communication*: The vehilcles exchange messages with neighbor vehicles within their geographical coverage. The exchange message with vehicles within their geographical coverage. In case of IoV, the nodes are not predefined in a particular geographical location, and they move from one location to another. The IoV can help to communicate the vehicles even outside of their geographical location.

(e) *Energy and processing capacity*: As compared with the IoT devices, the vehicles in IoV have an unlimited energy due to installation of huge battery power. The vehicles also have high processing capacity and memory space that can process complex computations. The IoV can handle the issues related with energy consumption and processing capabilities.

7.2.1 IoV Network Model

The network model of IoV can be categorized into three elements. They are cloud network, communication networks, and vehicles [3]. Figure 7.2 shows the network model of IoV with detailed components of each network elements. The brief description of each element of the IoV network model is discussed below.

7.2.1.1 Cloud Network

The cloud is one of the important network elements in design and development of the IoV. All the vehicles should connect and share information using the cloud through

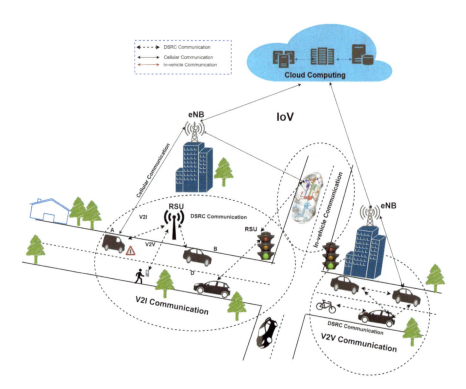

Fig. 7.2 IoV network model

Internet connectivity. The intelligent cloud infrastructure provides a range of services to the vehicles on the cloud platform such as basic cloud services, smart servers for applications, and information exchange services. With the increasing number of smart vehicles and various sensor devices installed in the smart vehicles, the traffic information collection, processing, and dissemination in real time will reach petabyte scale. The intelligent cloud computing can efficiently handle these type of ITS requirements. The clouds are used for operations such as uploading traffic information, processing the received information, and storing and distributing the information to the respective nodes. The smart application servers manage and process the traffic safety information, subscription services, and infotainment services. The servers are responsible for ITS information collection and end-to-end data delivery services to the vehicles. Some kind of artificial intelligence is used for making real-time intelligent decision for the smart vehicles based on gathered big data traffic information.

7.2.1.2 Communication Network

The communication network is the second network element that securely and reliably connects to the cloud and the smart vehicles. The communication network acts as a bridge to connect the intelligent cloud network and the smart vehicles for processing ITS-related services in IoV. There are different wireless access technologies to establish connection for efficient traffic information exchange. The vehicular communication is composed of heterogeneous networks based on various access technologies for specific type of connection as discussed in the previous chapters. In IoV, the different types of vehicular communication such as V2V, V2I, V2N, and V2P based on different access technologies (like Wi-Fi, cellular network, Bluetooth, etc.) play an important role for efficient communication. In IoV, due to the heterogeneous network environments, various access technologies are used for communication. The connection of the smart vehicles to the cloud should be efficient. In IoV, there should be seamless transfer of ongoing connection between different network operators, access networks, or heterogeneous networks. If the vehicles are registered with different cellular networks, the roaming module will perform seamless roaming procedure that prevent from network disconnection.

7.2.1.3 Vehicles

The vehicles, personal devices, and RSUs are the main elements in IoV that connect with the cloud networks through different communication networks based on various wireless access technologies. Each vehicles and other network devices has different network preferences and service requirements. In IoV, the vehicles and other devices utilize the smart cloud services depending on the application requirements. The different types of applications in IoV are safety, traffic management, and commercial based. The ITS applications such as safety, real-time traffic information management, navigation, parking are processed in smart cloud infrastructure, and the results are feedback to the vehicles accordingly. The IoV commercial applications such as insurance, carsharing, infotainment, and other cloud services are based on statistics managed by smart cloud that optimize the operation, reduce the cost, enhance productivity as well as improve the travel experience. The commercial applications incur some additional charges while providing enhance quality of services.

7.2.2 IoV Layered Architecture

The previous IoV architecture lacks security requirements such as authentication, authorization, and trust-related issues. They did not discuss the integration of network intelligence for selecting best radio access technology. There exists congestion and scalability issue due to the increase in number of vehicles as well as the absence of preprocessing of collected information from various sensors and neighbor vehicles. Gandotra et al. [4] proposed a three-layer architecture for IoV based on device-to-device communication. The first layer is the network layer where the devices are connected directly with each other or via network gateways. The second layer is the network management layer that manages connectivity of the devices by selecting appropriate wired or wireless network. It also provides unique IP addresses and roaming services so that the devices are always connected with each other. The third layer is the device-to-device (D2D) application layer that contains specific application, which supports IoV and other smart devices in smart city. Bonomi et al. proposed [5] a four-layered architecture considering the vehicular networks. The first layer is the end-to-end device that consists of vehicles, which communicate with adjacent vehicles using vehicular communication such as V2V communication. The second layer is the infrastructure layer that provides communication platforms to the lower layer vehicles such as V2I communication. The third layer is the operation layer that is responsible for managing and verifying information and regulate policies. The fourth layer is the service layer that provides different types of on-demand cloud-based services such as voice, data, and multimedia. Similarly, the authors in [3] add one more layer to the four-layer architecture and proposed five-layer architecture for IoV. The five layers are perception layer, coordination layer, artificial intelligence layer, application layer, and business layer. The perception layer is responsible for collecting data from several vehicular elements integrated with sensors and actuators. The coordination layer is responsible for coordinating different heterogeneous networks such as Wi-Fi, cellular networks, 802.11p, and short-range wireless networks. The artificial intelligence layer represents the big data analysis, processing, and storing in the cloud system. It is responsible for intelligent decision making on selection of the best application for intelligent services, traffic information, and infotainment as such. The application layer represents the smart applications aimed at end vehicle devices and performs intelligent services designed for vehicular networks

7.2 Internet of Vehicles (IoV)

in IoV. The fifth layer, i.e., the business layer represents the strategies for business models and investment designs based on statistical analysis using graph, flowchart, and comparison tables regarding data and budget pricing.

Castillo et al. proposed [6] a seven-layered architecture model for IoV that represents actual procedures and functions of each layer in depth as show in Fig. 7.3. We will discuss in depth regarding each layer of IoV architecture.

(a) *Interface layer*: The first layer is the user interaction layer that interacts directly with the vehicular nodes. The interaction layer is responsible for managing the notifications obtained from different interfaces such as auditory interface (like emitting alert/beep sound), visual interface (like flicker lights on windshield), and haptic interface (like vibrations on seat). It also manages the in-vehicle, inter-vehicle, and other different object interfaces.

(b) *Data acquisition layer*: The second layer is the data acquisition layer that collects information from different sources such as in-vehicle, inter-vehicle, sensors,

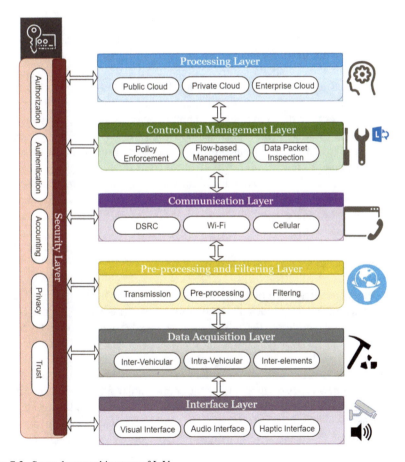

Fig. 7.3 Seven-layer architecture of IoV

actuators, RSUs, traffic lights, and other smart devices that are relevant to road safety, traffic data, and infotainment. As the vehicles are getting more smart and intelligent, a large number of sensors are installed in the vehicles. The vehicular communication is also getting efficient by the use of 5G and new radio (NR) access technology in cellular networks. Moreover, the advancement in 802.11p wireless networks generates huge amount of data. Hence, it is a challenging issue to gather, maintain, and disseminate all the information in real time.

(c) *Data preprocessing and filtering layer*: Only acquisition of various types of data is not enough it should be preprocessed and filtered. In IoV, efficient and proper aggregation of data, filtering, preprocessing, and management of those data is also very important. The irrelevant information from the huge amount of collected data should be preprocessed and filtered to reduce the data congestion, bandwidth and avoid the transmission of unwanted information in the network. An efficient and intelligent data mining technique can extract the relevant information from the collected data quickly, easily, and accurately. The transmission of data is based on the services subscribed by the vehicle.

(d) *Communication layer*: In IoV, there are several types of heterogeneous vehicular communication modes such as 802.11p, Wi-Fi, cellular networks, short-range wireless communications. Depending on the characteristics of communication environment, the IoV network provides seamless connectivity and services with optimal QoS to the users. The optimality depends on various parameters like available wireless technology, location and service requirements of the vehicles [7]. This layer provides an intelligent technique such as multiple attributes decision making (MADM) algorithms, which selects the most appropriate wireless network based on the appropriate information and measurement for finest QoS for the vehicles in IoV.

(e) *Management and control layer*: In IoV, a virtual network operator is used that is responsible for the management of various network service providers and controlling the data packet flow. This layer is also responsible for policy enforcement such as traffic management, data inspection, and traffic engineering. It provides a function to efficiently manage the information generated by all the various types of sensors installed inside the vehicles, neighbor vehicles, RSUs, etc.

(f) *Processing layer*: The processing layer processes the huge amount of information using different types of cloud computing infrastructures remotely and locally. This layer stores the information received from the lower layers and makes intelligent decisions based on the statistical analysis. It also identifies strategies for business models and investment designs sending the processed information to other services providers (SP). The SP reuses the information to develop new types of applications based on statistical analysis using graph, flowchart, and comparison tables based on the collected data and budget pricing.

(g) *Security layer*: The security layer is the last layer that communicates with all the other layers. It is responsible for the security of all the layers. This layer provides security by implementing security functions such as data authorization, authentication, accounting, access control availability, integrity, non-repudiation, confidentiality, privacy, and trust to exchange data among actuators and sensors through secure networks and service providers. This layer protects the IoV from different types of cyber-attacks and provides solutions to mitigate security issues.

7.2.3 Security in IoV

There is always a security threat in new and evolving networks. The new security standards and changing technology make it even worse. The IoV needs serious attention regarding security threats due to the involvement of human lives. A minor blunder can cause security failure that may result in severe accidents and takes life of the driver as well as cause damage to the vehicles. The security in IoV needs to be studied carefully to prevent it from serious attacks.

The security in IoV is of utmost importance that protects the vehicles from potential hackers and attackers. As the vehicles are connected to the Internet, the vehicles are prone to cyber-attacks that compromise the safety of the vehicles and the driver that might lead to severe accidents. Several security schemes have been proposed to secure the vehicular communication. However, the security researchers are successful in hacking the control system of Jeep Cherokee using the wireless connectivity of the vehicle's entertainment system [8]. They compromised the CAN bus control functions such as braking systems and steering by using the wireless communications (like Bluetooth and Wi-Fi) as attack vector. Similarly, the team from Tencent, a Chinese company, was able to hack the Tesla car by intercepting the Wi-Fi connection. They were successful in remotely controlling and activating the vehicle's braking system. This shows that the security systems in the vehicles have not be properly implemented that might cause vulnerabilities in the vehicle control system. Therefore, it is crucial to secure and protect the IoV system from vulnerabilities that sends critical system information. We will discuss the common IoV security attacks and security requirements that can prevent the vehicles from attackers.

7.2.4 IoV Security Requirements and Attacks

Because of the characteristics of IoV, it has several security requirements that need to be resolved before it is commercialized in the market. The list of security requirements of the IoV is as follows:

(a) *Authentication or non-repudiation*: The vehicle authentication and data authentication are one of the important requirements of IoV. When the data is transferred between the two vehicles, the data as well as the identity of the vehicles must be verified and authenticated. The authentication should be anonymous so that the unauthorized vehicles cannot gain access control to the vehicular system and hence prevent the system from unauthorized hackers.

(b) *Availability*: The availability in IoV ensures that the communication between the vehicles is available in different environmental conditions. There are different wireless communication modes, and the availability features help the vehicle to provide seamless connection to other vehicles. Therefore, they can communicate with each other and infrastructure anytime, anywhere even under high mobility conditions.

(c) *Integrity*: The integrity of the data in IoV ensures the consistency of data as well as the integrity to verify the information that has not been modified or changed by adding malicious content or data tampering. The sent and received information between vehicles must be checked properly to ensure that the data is transmitted correctly. This provides integrity to the exchanged message between two vehicles.

(d) *Privacy*: Privacy is one of the concerned issues in the IoV security. The IoV system should verify the data and the identity of the vehicles while providing privacy to the vehicle's private data. The exchanged information between vehicles and other entities must be protected to ensure privacy of the vehicle in IoV.

(e) *Access control*: The vehicles in IoV should only access existing services that they are entitled to access. The access control ensures reliability and security in the IoV system by assigning accountabilities to roles depending on the vehicles. The roles are assigned to the authentic vehicles of the IoV system.

There are several types of attacks in IoV. These attacks might compromise IoV system thereby affecting the stability and robustness of the system. The attacks might halt the IoV system and cause severe collisions. In IoV, the attacks can be categorized based on the different types of security requirements. The IoV security attacks corresponding to the security requirements are discussed below.

(a) *Authentication attacks*: In IoV system, several types of attacks hamper the authentication of the vehicles. Some of the authentication attacks such as Sybil attacks, masquerading attacks, and node impersonation attacks create multiple virtual identities for a single vehicle, or the malicious vehicles pretend to have the identity of other vehicles. This creates difficulty in identifying the information received by the receiving vehicle from the real vehicle. As a result, there is a chaos in making traffic decision based on received messages from multiple virtual vehicle identities available at a particular position. Some of the solutions to these attacks are identity-based cryptography (IBC), digital signatures, tamper-proof device (TPD), multiplicative secret sharing (MSS) technique, etc.

(b) *Availability attacks*: In this type of attacks, the attackers disturb the availability of the communication system or make the resources unavailable to the vehicles. Examples of availability attacks are channel interference attack, denial-of-service (DoS) attack, distributed DoS (DDoS) attack, greedy attack, etc.[9]. In case of channel interference attack, it interrupts the wireless communication between the vehicles or infrastructures. A jamming attack can be considered as channel interference attack that jams the communication signals so that the vehicular nodes cannot communicate with each other. In DoS attack, the attacker floods the IoV resources and consequently the vehicular elements are unable to handle as well as cannot further process the system request. The DoS attack creates severe damage to the IoV system, as the vehicular elements cannot process the real-time traffic information. Similarly, in case of DDoS attack, the attackers utilize multiple systems to attack a particular target system in a distributed manner. This attack floods the target system resources so that the target resources are unavailable to the vehicular nodes.

(c) *Integrity attacks*: In the IoV system, the integrity attack occurs when the malicious vehicles add false information or modify the original information exchanged between the vehicular entities. Some of the examples of integrity attacks are man-in-the-middle attack (MITM), session hijacking attacks, trust attacks, illusion attack, timing attack, message suppression attacks, etc. In MITM, the attacker sits in between the two vehicular elements that are exchanging information. The attacker secretly relays or modifies the exchanged information between two vehicular elements, who conceives that they are directly interacting with each other. In session hijacking attack, the attacker hijacks the communication session to gain unauthorized access and modify the information. In trust attack, the attacker exploits the trust relationship between different vehicular elements in the IoV networks.

(d) *Privacy attacks*: The vehicles must be able maintain their privacy at any cost because if the identity of the vehicle is revealed, it finally leads to the identity of the vehicle owner. The attacker might gain control over the vehicle as well as the vehicle owner that might cause severe human and economic damage. One of the privacy attacks is eavesdropping attack that is very hard to track, as it is a passive attack that does not interrupt any IoV network elements. It sniffs the exchanged messages between the vehicular elements targeting the privacy and finally gaining the passwords, access codes, and other information of the vehicles.

(e) *Access control attacks:* In IoV, each vehicle must be able to access authorized services only such as Internet, messaging, and collaborative network games. The hacker gains the access control and steals vehicles' credentials to impersonate a legitimate vehicle and carry out malicious activities. Few examples of access control attacks are dictionary attack, brute force attacks, etc. Some of the solutions to prevent this type of attacks are restricting access to the system, robust password policy, account lock policy upon unknown access, etc.

7.2.5 Challenges in IoV

The IoV system has to face different types of hurdles before it is adopted. All the critical issues need to be resolved before it is successfully adopted in the autonomous vehicle market. Some of the selected critical challenges in IoV are listed below.

(a) *Delay Constraints*: In IoV applications domain, the delay plays a very crucial role in delivering the safety-related messages. It requires a very strict delay constraint, where the delay should be maintained extremely low at any cost. During emergency, real-time exchange of critical messages is of utmost importance. It is difficult to make such an efficient IoV network with the existing communication infrastructure.
(b) *Lack of standards*: A proper communication standards in IoV system need to be developed to achieve a seamless communication and information exchange environment that allows transparent integration with existing standards. The lack of proper and open standard creates problem in further progress, improvement, and development of IoV system. An effective system can be made by integrating different communication systems for seamless information exchange.
(c) *Network connectivity*: In remote areas and countryside, there still exists poor and unstable network connectivity. The network connectivity is the backbone of the IoV system and this type of challenges in network connectivity should be resolved and improved as soon as possible. An intelligent and sustainable network connectivity can be designed to overcome this challenge.
(d) *Fault tolerance*: The IoV system must be fault tolerant, and the communication network should be highly reliable that provides a real-time communication even in the presence of malicious vehicles.
(e) *Interoperability*: Interoperability is one of the pivotal challenging issues for the interconnection of vehicles in IoV system. There are several interoperability challenges in IoV that must be addressed due to heterogeneous network models such as handoff timing and selection of the optimal wireless network technology to exchange information. The IoV should maintain well-organized and scalable management and communication among vehicular nodes.
(f) *Security and privacy*: Security and privacy in IoV system is of paramount importance as any system disaster directly affects the user safety. Security and privacy is one of the main challenging issues without which it is difficult to implement the IoV system. In the past, there has be several issues related to the security of the vehicular communications. The malicious vehicles breach the security and take control of the legitimate vehicles. In such case, the identification of malicious vehicle is required to be under surveillance and need to be traced so that the malicious vehicle can be identified and take necessary action against it. Privacy is equally important in IoV because if the information related to the vehicle is revealed to the hacker then the hacker might further disclose the information of the vehicle user and then finally obtain all the information of the vehicle.

7.2.6 IoV Applications

As the IoV is a special case of IoT, which is developed specially for automotive vehicles, several applications of IoV are indispensable for vehicles, drivers, pedestrian, smart city infrastructure, etc. The IoV applications are not only limited to vehicle convenience service, safe driving, traffic information, and crash response, but also it has a wide variety of application. The IoV applications can be classified into three broad types, viz. a. intelligent transportation system (ITS) application, b. business-related applications, and c. smart city-based applications. The broad classification of IoV applications is presented in Fig. 7.4. Numerous literatures are mostly focused on the IoV applications for ITS such as traffic efficiency, safety, comfort, and entertainment [10].

(a) *Intelligent transportation system (ITS) application*: The IoV for ITS is again categorized into four subcategories. We will briefly discuss some of them in this section. The safety-based IoV applications detect possible accidents on the road and prevent the vehicles from imminent collisions. A collision avoidance system (CAS) is used in ITS to disseminate safety-related information to the adjacent vehicles in IoV system. Some of the examples of safety applications are lane change warning, overtaking warning, automatic speed control, and automatic braking system. The navigation application within ITS is used to provide real-time information of the vehicles, location of the parked car, localization of events based on particular regions, and location sharing and tracking of the vehicles of the family members. The IoV applications for efficient ITS system are vehicle-related service information, optimal fuel usage information, carsharing application for less congestion on the road, and reducing pollution. The infotainment applications such as live multimedia streaming, downloading services, integration of vehicular networks with social network services such as Facebook and Twitter, and all time in-vehicle Internet connectivity provide comfort and leisure time to the driver and other passengers.

(b) *Business-related applications*: Similarly, there are some research done on the IoV applications for business and other services. The business-oriented IoV applications focus on creating business opportunities in IoV environment.

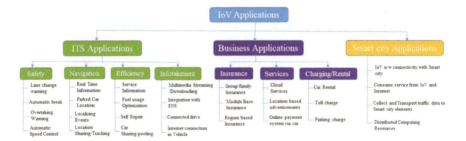

Fig. 7.4 Types of IoV applications

The new business models for different types of insurance can attract more customers and hence increase the revenue for vehicle-related insurance companies. There are different paid cloud-based services and location-based advertisement services that generate large amounts of money and increase economic values in the IoV system. The IoV can generate more economic values by taking advantages of the vehicle rental and charging services. The rental or charging services can generate more revenues by collecting the toll charges, parking charges, and other vehicle rental services. In addition, the electric vehicle (EV) bidirectional charging generates revenues by exchanging information in real-time regarding energy prices, on-demand energy sharing, and power loads. The integration of the wireless communication such as 5G and vehicular networks can satisfy the requirements of EV functionality and can reduce the fuel consumption, emission of harmful gases as well as increase efficient energy conservation by scheduling and reducing the peak load. The schedule charging and discharging scheme provides efficient energy management framework for smart grid and EVs [11]. The EV bidirectional energy exchange with vehicle to grid (V2G), vehicle to vehicle (V2V) and vehicle to home (V2H) makes the EV energy providers as well as energy consumers.

(c) *Smart City-based applications:* However, few researches are carried out regarding the IoV applications for smart city. The smart city applications open another dimension for IoV system. The IoV can be used to integrate with smart city to fulfill the requirements of smart cities by gathering huge data related to transportation, location, vehicle mobility, driver behavior and enhance the smart city surroundings. The smart vehicles do not have any battery, energy, or information processing constraints so they can be used as distributed nodes for processing and storing the information. The vehicles in IoV can be used as smart and intelligent nodes in smart city environment. The vehicles have seamless connectivity within the IoV networks, and they can sense and collect different types of information based on the inbuilt sensors while driving through the smart city. These vehicles gather huge amount of related data and send those data to the data centers within the smart city. The smart vehicles can act as additional distributed mobile resource nodes to supplement the constrained information processing resources for the smaller devices in smart city. The various sensors and OBUs on smart vehicles are used as data mules during the data gathering procedure, sends those data to the RSU, and then to the centralized smart city server for further processing. As different elements of IoV such as vehicles, RSUs, OBUs, and sensors are used, and the network overhead of the smart city can be managed efficiently. In addition, the historical paths of smart vehicles can be used to recognize the patterns in smart city and this can minimize the delay while ensuring satisfactory coverage. The mobility patterns of the intelligent vehicles and the in-built sensors of the vehicles can be used as Vehicular Sensor Networks (VSN) that measures and monitors the air quality of smart city environment. Moreover, in smart city, the smart vehicles' cameras and other sensors are used to monitor the urban location. The multimedia output can be used as a forensic evidence during accidents or other serious mishaps.

7.3 Machine Learning in Vehicular Networks

The intelligent and autonomous vehicles are enriched with multiple sensors such as LiDAR, radar, cameras that help vehicles to monitor their status in real time as well as perceive the surrounding vehicular environment. The installation of large number of sensors in and outside of the vehicles provides better sensing capabilities and transforms the vehicle from simple automotive mechanical device to powerful and intelligent computing vehicles. As we know, great power comes with great responsibility, similarly, the large number of sensors keeps generating, collecting, storing, and communicating large number of information to make every day autonomous vehicle greener, safer, and efficient. However, the large volume of information requires further filtering and processing to provide rich and real-time information regarding smart transportation system such as road condition, accident information, traffic environment, speed and direction of vehicles. The existing vehicular networks cannot process and utilize such abundant information to improve network performance.

The machine learning technique (MLT) stands as a strong and appropriate solution to vehicular networks' issues such as dynamic mobility, dynamic vehicle density, changing network topology, heterogeneous communication networks, and delay and real-time information [12]. There are different types of machine learning techniques and they are supervised, unsupervised, semi-supervised, reinforcement learning, and deep learning techniques as shown in Fig. 7.5. They exploit the information generated and stored in the network like vehicle's behavior patterns, locations, mobility, and network topologies to learn and extract the features. This provides efficient services like real-time traffic flow prediction and control, intelligent driving, location services, etc. A brief description of different types of MLTs are discussed below.

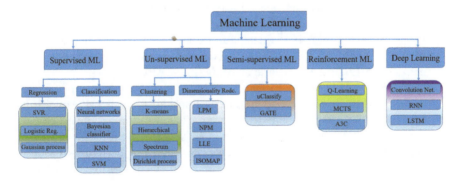

Fig. 7.5 Types of ML in vehicular networks

7.3.1 Types of Machine Learning Techniques

(a) Supervised learning: The supervised machine learning is based on direct supervision of the operation and its objective is to measure the data range and make forecasts of unknown, unavailable and future data based on labeled sample data. The final goal of the supervised learning is to map from input feature space to the labeled space to obtain the desired output [13]. The more the training data, the less is the error rate. It is further divided into two types, i.e., regression and classification. In regression algorithm, it identifies the patterns and computes the forecasts of continuous outcomes. Examples of regression algorithm are support vector regression (SVR), logistic regression, random forecast, and the Gaussian process for regression. In case of classification algorithm, the input data is labeled based on past data samples and the algorithm is manually trained to identify certain types of objects and classify them accordingly that provides discrete outputs (like Yes or No). It can be used to detect the network malfunction in vehicular networks. Some of the examples of classification algorithms are neural networks, Bayesian classifiers, K-nearest neighbors (KNN), decision trees, and support vector machine (SVM).

(b) Unsupervised learning: The unsupervised machine learning is based on unlabeled data that aims to find efficient representation of the data samples (extracting useful perceptions and detecting patterns) that are explained by hidden variables, i.e., the desired results are unknown and need to be defined. It is again divided into two types, i.e., clustering and dimensionality reduction. In clustering, the unlabeled data samples are grouped into different clusters depending upon their similarities. Some of the examples of clustering are k-means, hierarchical clustering, spectrum clustering, and the Dirichlet process. In case of dimensionality reduction, the algorithm reduces the high dimension to low dimension by removing unwanted noise from incoming data without losing relevant information. Examples of dimensionality reduction are linear projection method (LPM), nonlinear projection methods (NPM), local linear embedding (LLE), and isometric mapping (ISOMAP).

(c) Semi-supervised learning: The semi-supervised machine learning method lies in between the supervised and unsupervised machine learning. It exploits the limited set of labeled sample data to train itself with limited unlabeled data (pseudo-labeled) and then combines the labeled and pseudo-labeled data to create an algorithm that represents the characteristics of supervised and unsupervised learning. The semi-supervised learning is based on classification method to detect the data asset and then clustering method to group it into definite parts. It can be used in health care, legal, image, and speech analysis. Few examples of semi-supervised learning are uClassify and General Architecture for Text Engineering (GATE).

(d) Reinforcement learning: The reinforcement learning is based on exploration and exploitation technique with connected sequence of trial and error of incoming data with reward function that develops a self-sustained intelligence system. In reinforcement learning, at first, the action takes place and based on the action, the consequences are observed. Then, next action takes places considering the results of the previous action. A reward system is involved based on the particular sequence of action to drive the learning method. The objective of the system is to increase the positive rewards while decreasing the negative rewards. The nature of reward system may differ depending on the type of information. The common reinforcement learning are Q-learning, temporal difference (TD), Monte Carlo tree search (MCTS) and asynchronous actor-critic agents (A3C). In vehicular networks, the reinforcement learning develops optimal policies to meet diverse QoS requirements in dynamic wireless environment. The optimal policies are learned first followed by the actions taken by the vehicle nodes to adjust the power and allocate the channel based on dynamic environment.

(e) Deep learning: The deep learning is a subfield of machine learning technique based on artificial neural network that uses multiple layers to extract high-level features from the input data. It resembles the neural networks that consist of multiple layers of neurons. The deep learning can be built in supervised, unsupervised, and reinforcement learning. It consists of three layers, i.e., input layers, hidden layers with sigmoid functions, and output layer. The neural networks ability increases as the number of nodes in each layer and the number of hidden layers increases but with drawbacks like requiring large training data and high computation resources. The characteristic of deep learning is that it requires huge number of input data and high computational power like GPUs. Some applications using deep networks are convolutional networks, recurrent neural networks (RNNs), and long short-term memory (LSTM).

7.3.2 Type of ML in Vehicular Networks

With the advancement in ML techniques, it can be used as a cybersecurity solution in the autonomous vehicular networks. There are increasing research interest in developing cybersecurity solutions based on ML and deep/reinforcement learning. Connected vehicles can generate and collect terabytes of operational and diagnostic automotive data and store comprehensive geographic location information. That can be a good resource for the automotive framework to build next-generation cybersecurity solutions. The classical cyber security system cannot efficiently identify all the inside threats. The ML will dramatically help in developing accurate, real-time profiling and anomaly detecting capabilities to identify and capture user-based threats from within the system. ML-based cybersecurity is the identification of anomalies, where the machine detects abnormal or anomalous behavior or patterns in the data stream.

The ML-based cybersecurity solutions for vehicular networks can be used for vehicle network security, situational awareness, and vehicle intelligence. In case of vehicle network security, the intrusion detection or anomaly detection can be identified and isolated based on the exchange of information between the various automotive components and system such as telemetry data exchanged between ECU and ADAS [14]. The ML and deep learning based on LSTM and CNN can be used for complex architectures, which are capable of providing robust detection using information from multiple, spatially, and temporally isolated data streams. It can be used for malware classification, attack simulation besides intrusion detection. Some of the anomalies in the incoming data stream can be detected by the following procedure:

- Rule-based learning can be detected in the data stream by validation of each measurement derived from previous knowledge;
- Cross-check parameters like speed, location through multisensor data streams;
- Analyze the data stream via a temporal sliding window to identify abnormal patterns.

The vehicle situational awareness is a kind of defensive strategies. In case of vehicle situational awareness, automotive cybersecurity can be a security solution of VANET communication networks in the automotive cloud and can be of critical importance in both threat assessment and response. In this case, the detection is more challenging as predictions must be made with imperfect information within a rapidly evolving environment. Some of the situational awareness solutions based on ML are channel estimation and topology estimation [14].

Sometimes, the previous mentioned defensive mechanisms might fail in certain situations. The vehicle intelligence is essential to prevent the vehicle form taking unreasonable decisions that might risk the life of the driver and the vehicle itself. In situations like GPS spoofing attack or attacker taking control of the vehicle, the attacker gives wrong coordinates to the vehicle to drive into the building. Here, the vehicle intelligence based on ML solution has some kind of perception of its surroundings. It can prevent from the accidents as the vehicle can distinguish between safe and unsafe road areas. Some examples of vehicle intelligence are computer vision based on CNN, recurrent deep neural network to estimate trajectories of the vehicles, traffic flow prediction, etc.

We will overview the major machine learning types and their use in vehicle cybersecurity as follows:

(i) Supervised ML: The classical ML can be used to detect intrusion detection in vehicular networks based on dataset of clean and anomalous CAN messages. The classification can be used to detect the network malfunction in vehicular networks. In addition, regression can be used to forecast network throughput and channel parameters.

(ii) Unsupervised ML: In vehicular networks, the neighbor nodes are clustered into different groups depending on their characteristics so that the members of the group can send messages to the cluster head, which is energy efficient. The cluster heads use dimensionality reduction algorithm to filter out unwanted information on the aggregated data streams from the ECU, which is again analyzed to detect anomalous behavior. The unsupervised ML can be used to

7.3 Machine Learning in Vehicular Networks

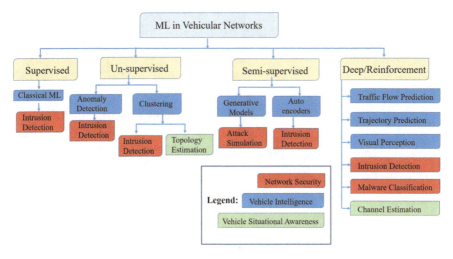

Fig. 7.6 Machine learning techniques in vehicular networks

detect anomaly detection as well as topology estimation used for situational awareness.

(iii) Semi-supervised ML: In semi-supervised learning, if the nonlinear relations in data streams with several hundred features are to be registered, deep generative models like auto-encoders (AEs) and variable auto encoders (VAEs) are needed. Using semi-supervised learning, security solutions such as attack simulations and intrusion detection can be achieved.

(iv) Deep/reinforcement learning: The deep learning and reinforcement learning provide a means of establishing autonomous cyber security so that human-related goals can be accomplished by taking human-like decisions. A reinforcement-based solution can interact in dynamic environment and they can manage resource allocation in changing vehicular environment to meet the diverse quality of service of vehicular networks. Some cybersecurity solutions based on deep reinforcement learning are given in Fig. 7.6.

7.3.3 Cybersecurity Solutions Based on ML in Vehicular Networks

In Chap. 3, we have discussed different types of attacks in intelligent vehicular networks and there are different solution methods to overcome those attacks. In vehicular networks, ML has not been extensively studied in detecting and solving various types of attacks. However, there are some significant works in vehicular network security based on ML such as misbehavior detection, DoS attacks, jamming attacks, and so on as shown in Fig. 7.7. We will discuss the most relevant MLT for security in vehicular networks.

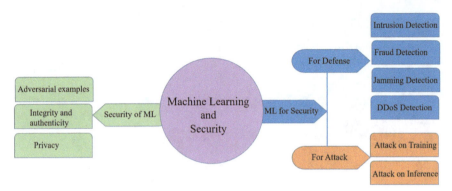

Fig. 7.7 Machine learning and security

7.3.3.1 ML to Detect Misbehavior Detection

The authors in [16] proposed a new type of misbehavior detection model using machine learning for vehicular networks applications. The misbehavior detection model is based on four phases, and they are information collection, information sharing, information analysis, and decision making. The detection model is evaluated by simulating a simulator called NGSIM, which is based on real-time traffic dataset. It consists of normal data traffic as well as malicious data traffic. The simulation dataset training is based on artificial neural network (ANN) that outputs very high accuracy in misbehavior detection.

7.3.3.2 ML to Detect DoS and DDoS Attacks

The authors in [17] used supervised machine learning called multi-class support vector machine (SVM) to secure software-defined vehicular networks (SDVN). This machine learning technique dynamically identifies several attacks such as probing attack, DoS attacks, user to root attack, and remote to local attack. The authors used MATLAB as a simulator that considers multi-class SVM toolbox. They used KDD CUP 1999 intrusion detection dataset to detect different types of attacks. The simulation results show that the SVM mechanism can detect and classify different types of attacks efficiently with high accuracy and decent performance.

Similarly, the authors in [18] used machine learning-based anomaly detection on SDN-based vehicular networks. The anomaly detection framework effectively and rapidly detects the DDoS attack in SDVN. The detection framework contains three models and they are detection trigger module, flow-table item collection module, and attack detection module. The detection trigger mechanism is based on the PACKET_IN message. The flow-table entry collection module is designed by merging the features of the DDoS attack and OpenFlow protocol to detect the DDoS attack. The attack detection module is based on SVM classification that is used to

7.3 Machine Learning in Vehicular Networks

train the samples and build a detection model to detect the DDoS attack in the vehicular networks. The DDoS attack traffic was generated using Scapy and hping3 for simulation, and the results show a very low false alarm rate.

7.3.3.3 ML to Detect Intrusion Attack

An intrusion detection system (IDS) based on deep neural network (DNN) for in-vehicle network security has been proposed for vehicular networks [19]. In this deep neural network, an unsupervised deep belief network is used to initialize the parameters as a preprocessing stage. In the in-vehicle network, the network packets and other information are exchanged between the ECUs. Then, the DNN is trained by multi-dimensional packet data to detect the fundamental statistical attributes of legitimate and malicious packets and excerpt the relevant features. Similarly, long short-term memory (LSTM) technique based on recurrent neural network (RNN) for CAN bus anomaly detection is used to detect intrusion attacks in vehicular networks [20]. The neural network is trained to forecast the subsequent packet data values. Its related errors are used as a signal for identifying anomalies in the sequence. The intrusion detection based on this technique detects and mitigates the malicious behavior on the CAN bus. The intrusion detection operates by learning to predict the next information originating from each sender on the CAN bus. The authors evaluate the detector by synthesizing anomalies with modified CAN bus data. The detector based on LSTM detects the manipulated and modified anomalies with low false alarm rate and high precision.

7.3.3.4 ML to Detect Smart Jamming Attack

The reinforcement machine learning technique can be used to detect smart jamming attack with the help of unmanned aerial vehicles (UAVs) [21]. The smart jammers are capable of discovering the OBUs and UAVs connection status, even convince the UAVs to use a particular relay strategy, and then attack it accordingly. The UAVs relay the OBU messages to nearby RSUs that has good radio transmission condition when the serving RSU is under jamming attack. This improves the communication performance between vehicular nodes. The processing of UAV message relay is a kind of dynamic game and the reinforcement learning based on Q-learning. It achieves the optimal strategy via trials and errors without knowing the strategy of the jamming model, if the game is long enough. There is a continuous interaction between the smart jammer and the UAV forming a dynamic game known as antijamming transmission game. The UAV selects its game strategy based on the system state (such as state of radio channel and BER of the OBU messages) and improves in each following states. The UAVs based on RL technique can decide whether to relay the OBU message to the nearby RSUs. The relay decision of the UAVs is based on the radio channel quality and BER of the OBU messages.

7.3.4 Attacks on Machine Learning/Deep Learning

Various types of attacks may occur in the ML or deep learning models. There are cybersecurity threats on ML such as computer vision based on CNN. CNN is vulnerable to adversarial inputs, which cause them to misclassify images such as classifying a car as a tree or a person as a traffic sign. When the attacker attacks, he applies different levels of noise to the camera's image output that are undetectable to the human eye before the system process and detects it. Thus, it is important to find CNN architectures that are reliable enough to protect the image data from adversarial attacks and use firewalls before creating cybersecurity solutions that depend on them. The attacks on ML can be occured in two ways as follows:

(i) Attacks during training time: In the first phase of ML, i.e., training phase, attacks may arise when training data is collected and fed to the ML model. If the user "contaminates" the training data intentionally by submitting inaccurate input data, it can cause the ML algorithm to malfunction or crash at the time of inference. To protect the attacker from modifying the ML data outputs, a standard application like anomaly detection can be used to train data sent by the user.
(ii) Attacks during inference time: In second phase of ML, i.e., inference phase, the privacy of the user must be maintained by protecting the user's private data. Let us consider an insider attacker situation where the user behaves maliciously. In this case, the user himself employs an attack called adversarial examples, where he feds legitimate input data to the ML model and the ML misinterprets the data. As a result, it causes serious concerns, such as misinterpreting the road signal during critical circumstances.

Machine learning proves to be a two-sided coin. On one side, the ML can be used for intrusion or malware detection application, while on the other side, the attackers can use ML techniques to enhance their attacking techniques. The attackers may concentrate on ML and AI to steal secret and critical information from secure systems, because it strengthens their learning capability. To protect ML from future adversaries, security by design and privacy by design concepts should be considered carefully from the beginning.

7.3.5 Application of Machine Learning in Vehicular Networks

The use of large number of sensors in the vehicular networks generates a huge data sources that can be collected, stored, and analyzed for decision making that is adaptive to dynamic vehicular environment. The machine learning techniques play an important role to provide artificial intelligence in vehicle and predict the unknown future events. It helps to provide safety to the vehicles by monitoring their status in

7.3 Machine Learning in Vehicular Networks

real time and perceive the surrounding vehicular environments. The large number of sensors provides improved detecting capabilities. There are several machine learning-based applications in vehicular networks that show significant advantages in user mobility prediction, handover optimization, localizations, etc. In vehicular networks, the above-mentioned machine learning techniques can be used to improve routing, networking, cache management, base station switching, user association, offloading, resource management, etc. In this section, we will discuss only task offloading and resource management based on machine learning in vehicular networks.

(a) Task offloading based on ML in vehicular networks:

In smart city, there will be large number of vehicles, pedestrians, and other network elements that generate huge amount of data and computation tasks. On the other hand, there are vehicles with high capacity and RSUs that provide their computing power and resources to the network. There are different types of communication mode used in V2X such as V2V, V2P, and V2I. The task offloading can be done in three different modes using machine learning based on the task collector and task originator. The three modes are V2V, V2I, and P2I2V offloading as shown in Fig. 7.8. We will briefly describe the three different offloading modes based on [15].

- (i) *V2V Offloading*: In V2V offloading, the distributed vehicles offload their computation task to the neighboring vehicles that has high and surplus computing power and resources. The individual vehicles that need to offload the tasks (i.e., V_T) discover the candidate service vehicles (i.e., V_S) that have high computing power within its communication range while moving in the same direction. In this mode, it is difficult to obtain the global information of

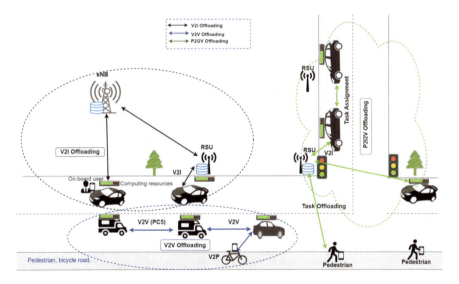

Fig. 7.8 Task offloading modes in vehicular networks

the adjacent vehicles such as available computing power to cooperate with each other.

(ii) *V2I Offloading*: In V2I offloading, the vehicles that need to offload the task select the infrastructures or RSUs with superior computing power in the absence of neighbor vehicles in its periphery. The infrastructures have high computing capability but the vehicles move with high mobility. The vehicles will have very short interaction with the infrastructures. This might affect the QoS provided by the infrastructures to the moving vehicles.

(iii) *P2I2V Offloading*: In this offloading method, the pedestrians on the road offload their computation tasks to other highly resourced vehicles by selecting the infrastructures like RSUs. The pedestrians select the nearby infrastructures as a medium to offload their task and the infrastructures send them to vehicles with suitable computing resources. The processed tasks are collected back by the infrastructures and then the tasks are feed back to the requested originator. In this offloading mode, the infrastructures have better global information regarding the neighbor vehicles but it has higher signaling overhead.

We will only describe the task offloading procedure based on V2V offloading mode using machine learning because of the limitation of the space. In V2V offloading, the candidate service vehicle (V_S) provides surplus services to the task vehicles (V_T) that is moving in the same direction as discussed in [15]. The V_T selects the best service vehicle V_S among the list of candidate service vehicles based on different parameters like delay, processing speed, location, velocity, etc., to offload its tasks. There are four phases to offload the task to the service vehicles and they are service vehicle discovery (SVD), task upload, task computation, and respond results.

(a) Service vehicle discovery (SVD): In SVD phase, the V_T search for the candidate service vehicle, V_S, from set of $N(t)$ at each time period t. It is assumed that $N(t)$ is dynamic due to the dynamic movements of the vehicles but it is not null.

(b) Task Upload: The task vehicle (V_T) selects one candidate vehicle n from set of $N(t)$ candidate vehicles to upload the tasks that need to be computed. The uplink transmission rate ($R_{t,n}^u$) between the V_T and V_S is given as

$$R_{t,n}^u = W\log_2\left(1 + \frac{H_{t,n}^u \cdot P}{\sigma^2 + I_{t,n}^u}\right) \quad (7.1)$$

where W is the channel width, P is the transmission power, $H_{t,n}^u$ is the uplink wireless channel, $I_{t,n}^u$ is the interference power at V_S, and σ^2 is the noise power.

However, there is a transmission delay incurred during the uplink process, which is given as

$$D_u(t, n) = \frac{x_t}{R_{t,n}^u} \quad (7.2)$$

7.3 Machine Learning in Vehicular Networks

where x_t is the input data size of the tasks in bits that needs to be processed.

(c) Task computation: During the task computation, the selected vehicle, V_S, processes the assigned task by the V_T with processing power w_t that is measured in CPU cycles per bit. The total workload to compute the task is given by $x_t w_t$. However, there is a computation delay (D_C) due to the request to compute multiple tasks at each time period t, which is given by

$$D_C(t, n) = \frac{x_t w_t}{f_{t,n}} \tag{7.3}$$

where $x_t w_t$ is the total workload and $f_{t,n}$ is the frequency of the CPU allocated to compute the task.

(d) Respond result: When the assigned offloading task is processed, the result is feedback from the candidate V_S to the original V_T. Similar to the uplink transmission rate, the downlink transmission rate ($R^d_{t,n}$) between the V_T and V_S is given as

$$R^d_{t,n} = W \log_2 \left(1 + \frac{H^d_{t,n} \cdot P}{\sigma^2 + I^d_{t,n}} \right) \tag{7.4}$$

where $H^d_{t,n}$ and $I^d_{t,n}$ are the downlink wireless channel state and the interference power at V_S.

Similarly, the transmission delay incurred during the downlink process is given as

$$D_d(t, n) = \frac{y_t}{R^d_{t,n}} \tag{7.5}$$

where y_t is the data size of the result responded by the V_S.

Hence, the total offloading delay $D_{total}(t, n)$, while offloading the task to the candidate vehicle V_T at time t is the sum of $D_u, D_C, and\, D_d$ given by,

$$D_{total}(t, n) = \frac{x_t}{R^u_{t,n}} + \frac{x_t w_t}{f_{t,n}} + \frac{y_t}{R^d_{t,n}} \tag{7.6}$$

As the task-offloading vehicle (V_T) lacks the global knowledge, while creating offloading decision, the authors in [15] designed the solution known as adaptive learning based task-offloading algorithm. This algorithm minimizes the expected average offloading delay given by

$$\min_{a1,\ldots,aT} \frac{1}{T}\mathbb{E}\left(\sum_{t=1}^{T} D_{total}(t, a_t)\right) \quad (7.7)$$

(b) Communication resource management based on ML

In V2X communication, the existing centralized resource management scheme based on central control incurs overhead while collecting information from all the vehicular nodes for making decision. The centralized resource management scheme does not work well as the number of vehicles increases in the network. A decentralized resource management scheme based on reinforcement learning and deep learning has been used in [16]. The deep reinforcement learning is a distributed resource allocation scheme that maps the partial observations of each vehicular nodes and helps in optimal resource allocation. It can deal with the latency requirements and optimization issue for V2X communication. This scheme ensures that there is minimum delay between the V2V communications as well as minimizes the interference between the V2I communications. The deep reinforcement learning focuses on V2V communication where the V2V agent interacts with the vehicular environment as shown in Fig. 7.9 [16]. The authors state that the vehicular network environment consists of several V2V connections that is outside of the considered V2V connections. The reason is that the other V2V connections cannot be controlled in the decentralized setting, and their actions like transmission power and selected spectrum are considered as the part of the environment.

In deep reinforcement learning, at time t, each V2V connection acts as an agent that observes a state (s_t) from state space (S). It takes action (a_t) that is selected from action space (A), which aggregates to choosing the subband and transmission power based on the decision policy (π). A Q function given by $Q(s_t, a_t, \theta)$ determines the decision policy (π), where θ is the Q function's parameter that is obtained using deep

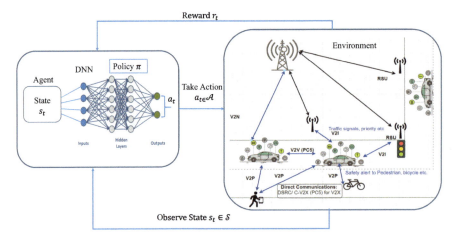

Fig. 7.9 V2V communication based on deep reinforcement learning

7.3 Machine Learning in Vehicular Networks

learning method. After taking the action (a_t), the environment transitioned to next state (s_{t+1}). A reward (r_t) is generated and obtained by the agent that is determined by the capacity of the V2V and V2I connections and their respective latency. The state observed by V2V connections for defining the environment is based on several parameters and the observed state is given as

$$s_t = \left[g_t,\ I_{t-1},\ h_t,\ B_{t-1},\ L_t,\ U_t\right] \tag{7.8}$$

where g_t is instantaneous channel information of V2V connection, h_t is V2I connection channel information, I_{t-1} is the previous connection interference, B_{t-1} is the election of neighbors previous time slot subbands, L_t is the remaining transmission load, and U_t is the latency constraints.

The training and testing data is generated from simulation setting based on 3GPP channel models that consists of V2V and V2I connections including their channel strengths. During the training phase, Q-learning based on deep neural network is used to get an optimal policy for V2V communication resource allocation to maximize the long-term collected discounted rewards (G_t). The optimal policy Q^* with Q-values is obtained without any information of the fundamental system dynamics. The data used for updating Q network is sampled from the memory, and the updated Q function [16] is given as

$$Q_{\text{new}}(s_t, a_t) = Q_{\text{old}}(s_t, a_t) + \alpha\left[r_{t+1} + \gamma \max_{s \in S} Q_{\text{old}}(s, a_t) - Q_{\text{old}}(s_t, a_t)\right] \tag{7.9}$$

where Q_{new} is the updated Q value and Q_{old} is the previous Q value.

The training samples generated for optimizing the neural network for deep reinforcement learning consists of s_t, s_{t+1}, a_t and r_t. Hence, the policy used in each V2V connection for choosing spectrum and power is stochastic at the beginning and gradually improved with the updated Q networks.

7.4 Vehicular Social Network

The vehicular social network (VSN) is the integration of social networks and the Internet of Vehicles (IoVs) that builds a social relationship among the vehicles as well as the drivers of the vehicles. It is also known as Social Internet of Vehicles (SIoV) [17]. In addition to social relationship, the VSN can combine the vehicular communication networks and human factors that influence the vehicular communication among the drivers of the vehicles. The mobility of the vehicles can establish the features of social networks where the vehicles show similar movements and routines while moving on the motorway. Some of the features like movement, channel condition, and driver's behavior features can improve the VSN's communication protocols and traffic-related services. One of the advantage of VSN is that it can overcome the issues faced by the IoV through integrating the features of IoV with social characteristics of vehicles and the drivers.

The VSN architecture consists of three layers, i.e., IoVs, VSNs, and social networks [18]. The VSN architecture is shown in Fig. 7.10, and it can be of three types, viz. centralized, decentralized, and hybrid architecture. The centralized architecture is based on V2I communication mode where the centralized server is responsible for managing and running the system. In case of decentralized architecture, the social networking and communication between the vehicles and other smart devices are carried out using V2V communication mode based on the DSRC communication. In case of hybrid architecture, the vehicles, drivers, and other smart devices communicate with each other based on V2V, V2I, V2N, and V2P communication mode. They select the communication mode opportunistically to connect with each other

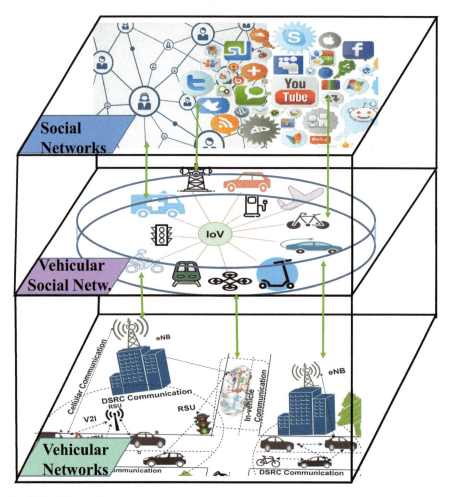

Fig. 7.10 VSN architecture

socially. In VSN, the spatiotemporal properties are very crucial for building social relationship between different vehicles and other smart devices on account to the dynamic nature of the vehicular networks. The vehicles and devices can communicate with each other when they encounter and are at close proximity with each other. The drivers of different vehicles around the same location can create a social network and communicate with neighbor vehicles concerning the local weather, road situations such as parking, traffic flow, traffic jam, and road accidents. They can create a dynamic community for a short period and discuss the relevant traffic issues at a particular time.

The VSNs have several features depending upon the movement of vehicles. One of the main features of VSNs is that the social relationship between the vehicles is dynamic because of the high movement of the vehicles. The vehicles can have social relation between each other based on their common interest even if they do not know each other previously. The social relationship will last until they are connected through wireless medium. The vehicles can have social relation with neighbor vehicles and smart devices on the fly based on specific content, specific location, and social relationship unlike online social network (OSN) that depends on family members and friends. Some of the other features of VSNs are heterogeneous connectivity, individual sociality, and limited bandwidth. One of the disadvantages of the VSN is that the lifetime of the social network among the vehicles is very limited, i.e., the social relationship is inactive when the vehicles leave the network. The basic characteristics are same for both the OSN and VSN but there are significant differences regarding the dynamic nature of vehicles, their interaction with society, network topologies, etc. As compared with the OSN, the VSNs are based on P2P communication where the social members are dynamic and the membership fluctuates any time based on location and interest with dynamic list of vehicular nodes as acquaintances. The OSN has higher security compared to the VSN due to the dynamic nature of vehicular networks. In addition, the social relationship lasts for unlimited lifetime in OSN. Nevertheless, it is possible to use the VSN in smart city due to its intrinsic features. The VSN will eventually interconnect every component of the smart city on different levels and creates new possibilities. However, the VSN introduces new challenges and endangers the network to new attack vectors. While the vehicles move in smart city, usually the vehicles follow the similar paths for their work and residence. The historical locomotive information along with the historical information of drivers' behavior and frequency of interaction among vehicles help to create distinct social profiles of the vehicles and the drivers, which in turn can be used to build vehicular social networks. There are new applications based on VSN's social networking concept such as RoadSpeak, Caravan Track, and VaniTweek. The RoadSpeak application [19] that allows the driver of the vehicles to exchange messages and voice chatting on the fly. In NaviTweet [20] and Caravan Track [21], these applications allow the drivers of the vehicles to share the mobility information among the group of vehicles interconnected with each other and can post the traffic updates on the social media platforms.

In VSN, the vehicles can create clusters by partitioning the network intelligently and form small vehicular groups that share similar characteristics, which is known as social clustering of vehicles [22]. The small clusters are more stable, where the cluster heads control and regulate the use of wireless channels, data aggregation, packet routing and scheduling. This helps in improving the bit error rate, high spectrum utilization, and gain throughput of the network, while decreasing the packet delays. The social behavior of the vehicles based on the above-mentioned characteristic helps to increase the stability of the social clustering of vehicles.

7.4.1 Applications of VSN

The VSN has several useful applications in the field of ITS that mainly focuses on the safety-related applications, infotainment applications, and distribution and transferring of multimedia services based on drivers' or passengers' interest. The data-driven applications are predominant in VSNs. The frequent mobility patterns of the vehicles help to induce a forecast about the future movement and trajectory information of the vehicles. The availability of the mobile computing systems in smart vehicles and devices generate and estimate the movement of vehicles in certain geographical locations. The applications of VSN are given in Fig. 7.11. The VSN applications can be categorized into three groups; they are social vehicular network applications, data-driven social networks, and social data-driven vehicular networks. Each of the categories further consists of different applications as shown in Fig. 7.11.

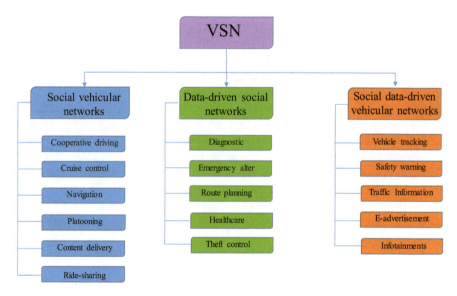

Fig. 7.11 Different applications of VSN

7.4.2 Security Issues

The VSN will be a part of smart city infrastructure that ultimately connects every component of the smart city on different levels and creates new possibilities. Along with the introduction of new technology, the VSN will face unavoidable new challenges and security issues. In VSN, the heterogeneous and distributed connectivity between vehicles in different geographical areas makes the security issues in VSN more complicated and difficult to resolve. In addition, VSN security not only includes the security related to the social characteristics of the vehicles but also the security and privacy issues of the drivers as well. The social aspects of VSN have increased the security demands that have a direct effect on the security of the overall vehicular networks. Some of the threats in VSN with specific emphasis on the social traits of the vehicles and drivers are described below.

(a) Malware: The malware is introduced in the network through the external software unit or the firmware updates. In VSN, the malware can spread in the network when there is software updates or firmware updates in vehicular components from other unknown or malicious vehicles. Malware such as worms or viruses can take control of the vehicles internal CAN system and interrupt the critical services.

(b) Sybil attack: In VSN, the Sybil attacks create several virtual identities for a single vehicle and pretend to have the identity of other vehicle. It is very difficult to identify the information received by the recipient vehicle from the real vehicle. There is a havoc in making decision based on received messages from multiple virtual vehicle identities available at a particular position. The existing solutions for vehicular networks cannot work for VSN. The Sybil attack along with the randomized attack patterns can compromise the VSN, which might lead to DoS attack against the wide smart city infrastructure.

(c) DoS attack: The DoS attack in VSN creates severe damage to the vehicular system, as the vehicular elements cannot process the real-time traffic information. The attacker floods the resources and the vehicular elements are unable to handle and process the system request that negatively affect the communication between the vehicular nodes and the infrastructures. The DoS attack might influence the large number of high powered computing facilities connected with the VSN in the smart city scenario.

(d) Impersonation attack: Similarly, the impersonation attack has severe impact on the VSN as the malicious vehicles steal the social identity of the neighbor vehicles or the drivers. The malicious vehicles then disseminate the security information under that vehicle's identity. This has a great influence on the decision of the other vehicle driver's decision and generate traffic problems.

7.4.3 Privacy Issues

The VSN is based on the characteristics of both the vehicular networks and the social networks of the drivers. The VSN has fundamental relationship between the vehicles, drivers, and other smart devices in the smart city. Therefore, we need to consider both types of privacy issues that arise in VSN. Location privacy is one of the important privacy issues in vehicular networks. The vehicles exchange critical messages between other vehicles and infrastructures using social networks that might contain the vehicles important information such as speed, position, and direction. This information can be linked with the vehicles as well as the driver's identity despite their strong privacy settings. The malicious vehicles might eavesdrop and obtain information about its social behaviors via wireless communication between the vehicles, and at the same time, it can deduce the location information through online social networks based on geo-tagging of messages. The malicious vehicles can easily track the location and their sensitive information and then infer the detailed social relationship and activities of the driver. The malicious vehicles might stalk, create digital record, or publish the private information in vehicular social network. In VSN, the privacy can be maintained by encrypting all the information that the vehicles disseminate or post in social networks. It is also necessary to protect the link privacy, i.e., the social relationship between other vehicles. In social networking, it is possible to deduce global relationship information by reverting a limited number of user accounts and identify the anonymous vehicles in social networks graph [23]. Thus, the objective of link privacy is to hide the type and existence of the social relationships.

7.5 Summary

In this chapter, we also discuss the vehicular social network (VSN). VSN is the integration of social networks and the IoV that constructs a social relationship between the vehicles and the drivers. The VSN is capable of social integration of the vehicular communication networks, human factors, and smart devices in autonomous ways. We present several useful applications of VSN concentrating on the safety-related, infotainment, and multimedia applications based on drivers' interest. The frequent mobility patterns of the vehicles help to induce a forecast about the future movement and trajectory information of the vehicles. While the emerging advanced mobility, such as IoV and VSN, provides a significant contribution in the transportation field, at the same time it possesses a variety of new types of threats. We discuss several types of machine learning techniques in IoV such as supervised, unsupervised, semi-supervised, reinforcement, and deep learning. We also discussed on how machine learning can be used as a cybersecurity solutions in vehicular networks. We provided few detailed applications of machine learning techniques in IoV. We also mentioned about the VSN applications and its security issues.

References

1. J.A. Stankovic, Research directions for the Internet of Things. IEEE Internet Things J. **1**(1), 3–9 (2014)
2. R. Shrestha, S. Kim, Integration of IoT with blockchain and homomorphic encryption: challenging issues and opportunities, in *Advances in Computers*, vol. 115, ed. by S. Kim, G. C. Deka, P. Zhang (Elsevier, 2019), pp. 293–331
3. O. Kaiwartya et al., Internet of vehicles: motivation, layered architecture, network model, challenges, and future aspects. IEEE Access **4**, 5356–5373 (2016)
4. P. Gandotra, R.K. Jha, S. Jain, A survey on device-to-device (D2D) communication: architecture and security issues. J. Netw. Comput. Appl. **78**, 9–29 (2017)
5. F. Bonomi, The smart and Connected Vehicle and the Internet of Things, in *Invited Talk, Workshop on Synchronization in Telecommunication Systems (WSTS)* (2013)
6. J. Contreras-Castillo, S. Zeadally, J. A. Guerrero Ibáñez, A seven-layered model architecture for Internet of Vehicles. J. Inf. Telecommun. **1**(1), 4–22 (2017)
7. A. Kumar, R.K. Mallik, R. Schober, A probabilistic approach to modeling users' network selection in the presence of heterogeneous wireless networks. IEEE Trans. Veh. Technol. **63**(7), 3331–3341 (2014)
8. A. Greenberg, Hackers remotely kill a jeep on the highway—with me in it, in *Wired*, 2015. (Online). Available: https://www.wired.com/2015/07/hackers-remotely-kill-jeep-highway/. Accessed 23 August 2019
9. A. Samad, S. Alam, M. Shuaib, M. U. Bokhari, Internet of Vehicles (IoV) Requirements, Attacks and Countermeasures, in *5 th International Conference on "Computing for Sustainable Global Development* (2018) pp. 4037–4040
10. M.R. Hafner, D. Cunningham, L. Caminiti, D. Del Vecchio, Cooperative collision avoidance at intersections: algorithms and experiments. IEEE Trans. Intell. Transp. Syst. **14**(3), 1162–1175 (2013)
11. X. Cheng, R. Zhang, L. Yang, *5G-Enabled Vehicular Communications and Networking* (Springer, Switzerland, 2019)
12. L. Liang, H. Ye, G.Y. Li, Toward intelligent vehicular networks: a machine learning framework. IEEE Internet Things J. **6**(1), 124–135 (2019)
13. N. Kaja, *Artificial Intelligence and Cybersecurity: Building an Automotive Cybersecurity Framework Using Machine Learning Algorithms* (The University of Michigan, Dearborn, 2019)
14. Z. El-Rewini, K. Sadatsharan, D.F. Selvaraj, S.J. Plathottam, P. Ranganathan, Cybersecurity challenges in vehicular communications. Veh. Commun. **23**, 100214 (2020)
15. L. Xiao, W. Zhuang, S. Zhou, C. Chen, *Learning- based VANET Communication and Security Techniques* (2019)
16. H. Ye, L. Liang, G.Y. Li, J. Kim, L. Lu, M. Wu, Machine learning for vehicular networks: recent advances and application examples. IEEE Veh. Technol. Mag. **13**(2), 94–101 (2018)
17. S. P. George, N. Wilson, N. U. Krishnapriya, M. Kareeshma, Social Internet of Vehicles. Int. Res. J. Eng. Technol. **4**(4), (2017)
18. Z. Ning, F. Xia, N. Ullah, X. Kong, X. Hu, Vehicular social networks: enabling smart mobility. IEEE Commun. Mag. **55**(5), 16–55 (2017)
19. S. Smaldone, L. Han, P. Shankar, L. Iftode, RoadSpeak: enabling voice chat on roadways using vehicular social networks, in *Proceedings of the 1st Workshop on Social Network Systems* (2008) pp. 43–48
20. W. Sha, D. Kwak, B. R. Badrinath, L. Iftode, Social vehicle navigation: integrating shared driving experience into vehicle navigation, in *Proceedings of the 14thWorkshop on Mobile Computing Systems and Applications* (2013) pp. 26–27
21. C. Squatriglia, Ford's Tweeting Car Embarks on American Journey 2.0, in *Wired*, (2010)
22. L. Maglaras, et al., "Social Internet of vehicles for smart cities. J. Sensors Actuator Netw. **5**(3) (2016)
23. A. Narayanan, V. Shmatikov, De-anonymizing Social Networks, in *Proceedings of the 2009 30th IEEE Symposium on Security and Privacy* (2009) pp. 173–187

Chapter 8
V2X Current Security Issues, Standards, Challenges, Use Cases, and Future Trends

8.1 Overview

The Fourth Industrial Revolution will see autonomous vehicles at the center. The automotive industry has already become a worldwide issue that is not constrained solely by one country's strategy. Moreover, autonomous vehicles are also fundamentally changing the behavior of human mobility. It will also bring in product creativity focused on those innovations. Therefore, the international technological infrastructure around this topic is already under demand to develop a new set of standards. New rules and laws are replacing existing ones, and new relevant legislations are being discussed worldwide. We are at the peak of a major breakthrough in vehicle communication and safety on the road. The vehicle-to-everything (V2X) technology has already been applied to improve road safety through its significant past and current developments. The vehicle-to-everything (V2X) works on 802.11p protocol. The V2X communication based on 802.11p does not offer preinstalled infrastructure and complete spatial connectivity as compared to cellular networks, but is a major driver for vehicle safety and traffic efficiency applications with its capabilities for direct communication among short communication networks. The V2X vehicle technology can save lives, but only if these technologies are granted opportunity to advance and implement in the society. The fast evolution and rise in the development of modern systems and networks, combined with the complexity of increasing risks, pose daunting challenges in preserving the autonomous vehicle security. The security solutions require a stable and safe network infrastructure, but they also need to protect the privacy of the drivers as well as the vehicles. The cybersecurity is very important in autonomous vehicle and its application because the life and property of the passenger are dependent on the autonomous vehicle security. It should be dynamic as the cyber-attack vectors are developing frequently and the autonomous vehicles should be protected from new cyberattacks. The security management should be proactive as well as reactive that are based on dynamic approach.

This chapter gives a brief overview on international cybersecurity regulations, standardizations, different types of organization working in DSRC and C-ITS protocols in Sect. 8.2. Section 8.3 discusses the V2X technology based on DSRC and cellular network and its adoption. Then, Sect. 8.4 discusses the 5G V2X testbed and its use cases. Finally, Sect. 8.5 presents the future of intelligent and autonomous vehicles and its cybersecurity issues.

8.2 Standards, Regulations, and Legal Issues

There are several organizations, consortiums, associations, and authorities. We will discuss only the major ones, which are given below:

8.2.1 International Cybersecurity Standardization in Automotive Industry

The international cybersecurity standard effort in automotive industry had already been discussed in Chap. 1. In this section, we will briefly discuss only the major international cyber security standards, organizations, consortiums, associations, and standards that have developed new standards and recommendations. They are categorized into EU, USA, and global standardization.

A. **Europe initiatives**: Regarding vehicle safety and security, there are several cybersecurity regulations, initiatives, and projects carried out by EU countries.

   1. EU Projects: In 2006, the SEcure VEhicle COMmunication (SEVECOM) project started to deal with security of vehicular communications and inter-vehicular communications. It provided solutions to the problem that are specific to the road traffic information. In 2008, the E-Safety Vehicle Intrusion protected Application (EVITA) project started and its goal was to design and verify OBU prototypes and provide e-safety by securing the electronic components of vehicles from tampering. The EVITA project secures inter- and intra-vehicle communication by allowing trustworthy communication between the vehicles and the internal components and the project ended in 2011. From 2008 to 2012, the SimTD project was carried out in Germany. Its objective was to increase road safety and improve the traffic efficiency based on V2X communication. The result of this project can be applied in the categories like traffic and value-added service. The 7th Framework Program of the EU commission started the Open VEhiculaR SEcurE (OVERSEE) project in 2010 and ended in 2012. OVERSEE provided standard, secure, generic communication application platform for vehicle and

8.2 Standards, Regulations, and Legal Issues

enhanced the efficiency and safety of the road traffic. Similarly, the framework funded another project called Preparing Secure Vehicle-to-X Communication Systems (PRESERVE) in 2011 and ended in 2015. The PRESERVE objective was to design an integrated V2X Security Architecture (VSA), implement the architecture, and field test the VSA system. Conversely, in 2012, the Bosch established its own hardware security module (HSM) specification to harden the embedded system like ECU that fulfills the vehicular requirements. In 2014, Intelligent Transport System Security (ISE) project was introduced in France, which contributed to security and privacy of future vehicle of C-ITS. The goal of ISE is to design and implement a proof of concept of ITS European PKI that has been proposed as a standard at ETSI [1]. SEcurity and SAfety MOdelling (SESAMO) is a European industry-based project by Advanced Research and Technology for EMbedded Intelligent Systems (i.e., EU-ARTEMIS project) that runs from 2013 to 2015. SESAMO objective is to increase the safety and security of the embedded system and products in various areas like ITS, industry, avionics, etc. The AI safeguarding project started from July 2019 and will be completed in June 2022 with a total budget of 41 M Euro. The goal of this project is to generate security, functional safety measures, and safeguarding strategies for the AI functions used in autonomous vehicles.

2. EU Consortium: The CAR2CAR Communication Consortium (C2C-CC) was established in Europe, which is a consortium of leading European and international vehicle manufacturers, equipment suppliers, engineering firms, road operators, and research institutions. The objective of this consortium is to save lives by investigating and developing C-ITS solutions that helps to achieve vision zero, i.e., to implement accident-free traffic by overcoming road accidents. Volkswagen, which is one of the members of C2C-CC, launched new vehicles VW Golf 8 that supports C-ITS that is based on ETSI ITS-G5 standards for establishing short-range communication between vehicles [2].

3. EU Standardization: The AUTomotive Open System Architecture (AUTOSAR) is very popular in-vehicle software standardization organization for intelligent and autonomous vehicles. The AUTOSAR is a global consortium of automakers, suppliers, service providers, vehicle industry, semiconductors, and software company. Their objective is to build a global open industry standard for vehicular software architecture. It specifies a message authentication function for identifying and thwarting falsification and spoofing of communication data from the AUTOSAR specifications. The AUTOSAR introduced a secure authentication mechanism for critical data on a PDU level called Secure On-Board Communication (SecOC), which describes the basic functionality, security features, and API of AUTOSAR SecOC module. Figure 8.1 shows the timeline of safety and cybersecurity standards, and projects in automotive vehicles.

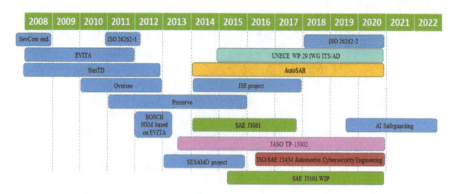

Fig. 8.1 Timeline or safety and cybersecurity standards, and projects in automotive industry

B. **North America Initiative**:

1. US Organizations: The Society of Automotive Engineers (SAE) is a global standard developing organization for engineers and professionals in different industries and associations [3]. It stipulates guidelines and the cybersecurity functional development process for vehicular attacks by introducing J3061 as defined in Cybersecurity Guidebook for Cyber-Physical Vehicle Systems in 2014. In 2015, the SAE introduced J3101 that defines a common set of requirements for hardware-protected security for terrestrial vehicle applications to expedite security in vehicular applications. The Society of Automotive Engineers (SAE) and International Organization for Standardization (ISO) jointly worked together to develop current state-of-the-art cybersecurity standards for vehicles in two areas, i.e., road vehicles and ITS. The SAE and ISO co-chaired and worked as a Joint Working Group (JWG) to introduce ISO/SAE 21434 under a new agreement. The SAE J3061 was based on cybersecurity of vehicle. The objective of ISO/SAE 21434 is to provide cybersecurity activities for all phases of vehicle life cycle, manage cybersecurity activities, and analyze risk factors for cybersecurity requirements.
2. US Standardization: In the USA, the Institute of Electrical and Electronics Engineers Standards Association (IEEE-SA), a part of the Institute of Electrical and Electronics Engineers (IEEE) organization, provides support, development, and advancement of global standards across a wide range of industries. It introduced 802.11p standard that is an extension of the Wi-Fi 802.11 standard in order to support Wireless Access in Vehicular Environments (WAVE) for ITS applications. The 5G Americas is an industry trade organization made up of major suppliers and telecommunications service providers. The mission of the organization is to support and promote the development of wireless technologies such as 5G and beyond, its services, applications, and smart devices in the Americas. The 5G Americas invests in the development of a connected wireless society while leading 5G development across all of America. The 5G Americas enhances vehicle cybersecurity

by increasing user privacy, network resilience to cyberattacks, and strengthening hardware protection for vehicles. This goal is accomplished a secure device credential provisioning and storage, and new network enhancements provide vehicle-to-network communication security.

C. **International Initiatives**: A group of experts from around the world assembled to develop international standards known as ITU's Telecommunication Standardization (ITU-T) that defines global infrastructure elements for ICT. The International Organization for Standardization (ISO) advocates proprietary, technological, and commercial practices worldwide. The ISO and SAE work together for the vehicle cybersecurity standards. The ISO 26262 is an international risk-based standard for functional safety of electronic and electrical systems in vehicles derived from IEC 61508 and defined by ISO. It is designed for functional safety of automotive electrical and electronic devices by guiding and regulating throughout the entire product life cycle process. It was first introduced in November 2011 to address the safety hazards due to the malfunction of electronic equipment of the passenger vehicles. However, the first edition was withdrawn and then it was revised and again republished second edition as ISO 26262-2 in 2018. One of the parts of second edition focused on road vehicle functional safety and standards for Safety of the Intended Functionality (SOTIF). It helps in avoiding system failures and detect random hardware failures.

1. United Nations Economic Commission for Europe (UNECE): UNECE is preparing a certification for a Cyber Security Management System (CSMS) that mandates the approval of the vehicles according to the requirement of the recent document proposal. The UNECE is working toward a global standardization and regulation on cybersecurity of the vehicles since December 2014. There is an increasing cyber-attack that risks the vehicular communication due to weak infrastructure for online updates of vehicles like over-the-air (OTA) issues [4]. To provide cybersecurity and data protection and solve OTA issues, a task force was created under Informal Working Group on Intelligent Transportation Systems/Automated Driving (IWG on ITS/AD) within WP.29. The UNECE WP.29 IWG ITS/AD document provides a guideline on cybersecurity, relevant standards, regulations, and data protection that are applicable to the automotive industry. The software update regulation contains requirements for OEM software update that include demands for safe and secure updates [4].
2. 5G Automotive Association (5GAA): 5GAA is a global, international cross-industry organization for automotive, telecommunications, and technology enterprises [5]. Before the introduction of 5GAA, different industries have their own definition, terminology, interest, and use of V2X cases in different parts of the globe, which make it difficult to communicate in one common language. The 5GAA solves this problem by creating a universal global language for the V2X under one umbrella across different industries. The goal of 5GAA is to develop, evaluate, implement standards, promote C-V2X solutions, and support 5G connectivity technologies that drive global

market penetration and usability, and facilitate standardization for C-V2X. The 5GAA signed a letter of intent for collaboration with the European Automotive Telecom Alliance (EATA) in 2017. It also works for standardization for driverless, autonomous vehicle in cooperation with ETSI, 3GPP, and SAE. In late 2019, the 5GAA collaborated with the ETSI for testing C-V2X for the first time in Europe. The tests executed with 95% success rate confirming a high level of multi-vendor interoperability. The successful test was an important milestone to support commercial deployment and global strategy to accelerate the C-V2X throughout the world with OBU including the infrastructure.

The 5GAA Working Group 7 (WG7) evaluates the currently available security solutions in V2X communication and recognizes gaps toward comprehensive, secure end-to-end solutions besides specifications. The 5GAA promotes the use of the entire 5.9 GHz spectrum for ITS safety applications. Strategic initiatives by governments, regulatory agencies, industries, and research institutions are also required on account of the ambiguity of communication systems, lack of support of networks, total cost of deployment, and exposure to certain technological issues, like enhancing cybersecurity. 5GAA acknowledges that C-V2X not only has the advantages of future proofing and the use of wireless connections in tandem, but also has better direct communication efficiency relative to 802.11p. It provides a higher degree of security across all operating modes such as embedded security for vehicle-to-network (V2N) transmission, and the equivalent public key infrastructure (PKI) security services defined in standards using 802.11p. The back-end V2N services manage delay-tolerant or wide area network applications and provide enhanced security and privacy from suitable network locations. The service providers may provide specialized services such as ADAS and CAD assistance with low latency requirements in network front-end infrastructures. Figure 8.2 shows the probable security framework for end-to-end terminals, network front and back ends [6].

The automotive cybersecurity testing, validation, integration, and standardization process are given in Fig. 8.3 [7].

Additional information on international cybersecurity standard effort in automotive industry in Asia and other parts of the world is given in Appendix A of this book.

8.2.2 Standardization for V2X Communication and Frequency Allocation

The international development of ITS has been the driving force for the advance deliberation of regulatory provisions for vehicular communication systems. Significant advances have been made in many regions, particularly in Europe, USA, and Japan.

8.2 Standards, Regulations, and Legal Issues

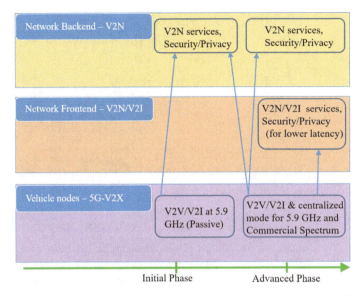

Fig. 8.2 Security framework for end-to-end terminals, network front and back ends [6]

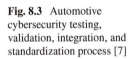

Fig. 8.3 Automotive cybersecurity testing, validation, integration, and standardization process [7]

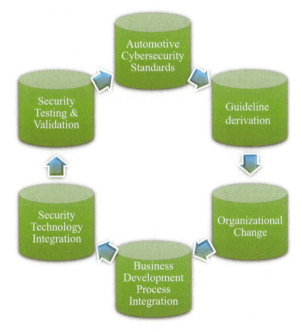

Table 8.1 ITS spectrum in different regions of the globe

Region	Country	Year	ITS spectrum (MHz)	
North America	USA	1999	5852-5925	
European Union	Europe	2008	2008/671/EC	5875-5905
			ECC/REC/(08)01	5855-5875
			ECC/DEC/(08)01	5875-5925
Asia Pacific	Japan		755.5-764.5 and 5770-5850	
	Korea	2016	5855-5925	
	China	2018	5905-5925	
	Singapore	2017	5855-5925	
Australia	Australia	2017	5855-5925	

International organizations, including ITU, European CEPT, United States Department of Traffic (USDoT), Federal Communications Commission (FCC), and Asia-Pacific Telecommunity, are envisioning spectrum harmonization initiatives. They are also beginning to consider the 5.9 GHz band ITSs in their entirety. While writing this chapter, the global ITS spectrum is in further review under ITU-R Working Party 5A, which is liable for land mobile service research, including wireless access and ITSs while excluding IMT.

A. **Europe**: In 2008, the Electronic Communications Committee (ECC) passed a recommendation (ECC/REC/(08)01) for the non-protected and non-interference frequency use at 5855-5875 MHz and a decision (ECC/DEC/(08)01) regarding ITS non-safety applications in the 5.9 GHz frequency band as given in Table 8.1. Again, in the same year, the ETSI allocated a 30 MHz band (i.e., 5875-5905 MHz) for time-critical road safety-related application of ITS and additional 20 MHz band for future ITS applications in 5.9 GHz frequency based on the European Commission Decision 2008/671/EC [8]. In 2014, ETSI along with European Committee for Standardization (CEN) released the first version of V2X technology known as ETSI ITS-G5. Since then, ETSI has been working toward improving the regulatory framework in radio interferences and cyber security. However, this recommendation might not accommodate the 5G technologies, for which adoptions of other bands such as the 63-64 GHz band may be reasonable. In 2018, EU mandated the use of V2V technology in all the new vehicles [9].

B. **USA**: In 1999, the US-DoT and FCC regulated the channel allocation of 75 MHz band in 5.9 GHz frequency spectrum, i.e., 5850-5925 MHz. It reserves one 10 MHz control channel band (5885-5895 MHz) for vehicle safety and another 10 MHz band (5915–5925 MHz) for public safety communication and rest for general use as specified in FCC 47 Code of Federal Regulations (CFR) Parts 0, 1, 2, 90, and 95 amendments for DSRC. In 2016, the FCC requests the proposal of Unlicensed National Information Infrastructure (U-NII-4) prototype units to show unlicensed devices and DSRC coexistence capabilities. In 2004, IEEE

8.2 Standards, Regulations, and Legal Issues

formed a taskforce to work for Wireless Access in Vehicular Environments (WAVE) that led to the IEEE 802.11p and then revised in 2010. In 2012, the IEEE 802.11p was combined with IEEE 1609 and SAE J2735 to allow a full-standardized message stack that was approved by DOT for DSRC applications. In 2014, the US National Highway Traffic Safety Administration (NHTSA) issued a report claiming that the V2X platform was technologically tested and available for real market implementation. The NHTSA mandates the use of V2V technology in all the new vehicles to accelerate the proposed schedule and minimize the collision-based incidents [9].

C. **Asia Pacific Region**: The vehicular communication standardization was also followed by other countries in Asia Pacific regions such as South Korea, Japan, Singapore, China, and Australia between 2016 and 2017. The 5.9 GHz band is the prospective band used for ITS applications in Asia Pacific region as shown in Table 8.1. In 2016, Korea allocated the 5855-5925 MHz band in 5.9 GHz frequency for C-ITS (V2V and V2I communications) based on a 10 MHz channelization by the Ministry of Science and ICT (MSIT) that is similar to EU and USA [10]. There is an ongoing amendment to the Korean ITS standards in the Telecommunications Technology Association (TTA), and one of its objectives is to support various radio technologies for ITS applications, including C-V2X along with IEEE 802.11p.

In October 2018, China introduced Intelligent Connected Vehicles (ICV) allocated 20 MHz band (i.e., 5905–5925 MHz) in the 5.9 GHz spectrum to be used as Internet of Vehicles (IoV) accommodating LTE-V2X technology. In Japan, the C-ITS implements two different spectra for ITS applications; they are 755.5–764.5 MHz band and 5770–5850 MHz band as shown in Table 8.1. In October 2017, the Telecommunications Standards Advisory Committee (TSAC) of Singapore proposed a specification for ITS and allocated the frequency spectrum from 5855 to 5925 MHz. The spectrum is divided into 10 MHz channel in 5.9 GHz for DSRC operation. The Australian Communications and Media Authority (ACMA), which is a communication and media regulatory body, allocated 5855–5925 MHz in 5.9 GHz band for ITS communication in 2017. A detailed ITS spectrum allocation in Asia Pacific is given in Appendix B.

8.2.2.1 International V2X Standardization

International standards are essential requirement for the deployment of V2X communication systems that provide requirements to enable connectivity between V2X systems and devices, along with interoperability of implementations from various vendors. However, standardization of V2X often poses multiple challenges. From technical point of view, the V2X standards contain a large number of reference and test requirements on various fields spanning from radio and protocols for safety and applications. The vast number indicates a huge complexity of incompleteness and ambiguity.

Table 8.2 Difference between DSRC and ITS-G5

Layers	Usage	DSRC (US)	ITS-G5 (EU)
Applications	Safety and traffic applications	Safety (e.g., VCA) and non-safety (e.g., traffic estimation)	RHS, ICRW, LCRW
Facilities	V2X message	BSM, event-driven message (EDM)	DENM, CAN
Networking and transport	Message protocol	WSMP	BTP, GeoNetworking, GN6
Access technologies	PHY/MAC	IEEE 802.11p	ITS-G5
Security	Module	IEEE 1609.2	ETSI TS 103 097

Developing release specification poses both the challenges of forward and backward compatibility between releases, especially with the addition of new features and categories of the framework. From a non-technical point of view, V2X standardization is discussed by several Standardization Development Organizations (SDOs), which have developed partially overlapping requirements. The integration of guidelines from various SDOs is daunting, and compatibility of the standards is time and resource consuming. Standardization of V2X communication began with the allocation of the 5.9 GHz frequencies in the USA that was issued in 2002. In recent decades, parallel standardizations have been established in the USA and Europe, primarily because the initiatives were funded by numerous technology development projects and sponsored by specific stakeholders; eventually, they contributed to a separate set of standards. These two standards will be referred to as dedicated short-range communications (DSRC) in the USA and cooperative intelligent transportation system (C-ITS) in Europe. The overall protocol stacks for both DSRC and C-ITS are similar, but there are some difference in each layer as shown in Table 8.2. There are two types of DSRC application layers, and they are safety and non-safety applications. The safety applications such as Vehicle Collision Avoidance (VCA) helps in reducing the road accidents by avoiding the vehicle collisions, while non-safety applications such as real-time traffic estimation and infotainment applications help to estimate the travel time and make the travel more pleasant. The applications in the ITS-G5 are not standardized directly. The applications are categorized into three types, and they are road hazard signaling (RHS), intersection collision risk warning (ICRW), and longitudinal collision risk warning (LCRW). The RHS signals are related to emergency brakes, emergency car approaching, and dangerous roads. The ICRW and LCRW signals are the crash hazards related to head-on accident or crash at road intersections.

The V2X message functionality at the DSRC facility layer consists of a basic safety message (BSM) and event-driven message (EDM). The BSM is a beacon message broadcasted every 100 ms periodically, and it consists of vehicle state information such as position and status. The BSM is usually broadcasted over a smaller area within 150–300 m radius. The EDM safety messages are transmitted when an event occurs

8.2 Standards, Regulations, and Legal Issues

on the road such as accidents, and they should be transmitted with low collision and short delays over a wider area. On the other hand, the V2X message functionality in ITS-G5 consists of two types of messages, i.e., Cooperative Awareness Message (CAM) and Distributed Environmental Notification Message (DENM). The CAM is based on ETSI EN 302 637-2 and is similar to the BSM of DSRC protocol. The CAM sends vehicle state information periodically for identifying the vehicle's mobility and location. The DENM is based on ETSI EN 302 637-3 that transmits traffic safety information to the neighbor vehicles in a particular geographical area only when there is some event.

The IEEE 1609.3 defines a protocol named Wave Short Message Protocol (WSMP) that is a single-hop network protocol with a low header of few bytes, and the message used by WSMP is WAVE Short Messages (WSM). By transmitting periodical messages known as WAVE Service Advertisements (WSA), the WAVE system advertises its services. Each WSA may list PSIDs for network services and information that is required for the receipt and processing of WSMs for each service being advertised. In ITS-G5, the ITS describes a multi-hop communication ad hoc protocol known as the GeoNetworking protocol defined in ETSI EN 302 636 standard. The feature of this protocol is to address and transmit geographical coordinates. In contrast to the WSMP, GeoNetworking is optimized for multi-hop communication with geo-addressing, offering additional technical support functionality, but at the cost of higher complexity and overhead of the protocol. In addition, this layer consists of adaptation sublayer called GN6, i.e., IPv6 over GeoNetworking.

The access technology layer for DSRC covering PHY and MAC operates in IEEE 802.11p 5.9 GHz band. The IEEE 802.11p equivalent in the EU C-ITS stack is known as ITS-G5 where the G5 refers to the 5 GHz frequency band. In ITS-G5, the MAC layer uses un-coordinated channel access scheme such as Enhanced Distributed Channel Access (EDCA), which is contention-based that uses the Carrier Sense Multiple Access with Collision Avoidance (CSMA/CA) for traffic data prioritization.

As for the security module, the DSRC uses IEEE 1609.2 standards for authentication and encryption of the messages using digital signatures and certificates. The USDOT mandates the use of vehicular PKI (VPKI) defined by 1602.9 standard. The VPKI will provide the vehicles with a series of pseudonymous provisional certificates that can be used to digitally sign V2X messages and to verify the reliability and integrity of the shared data for other vehicles. The pseudonymous certificates used by the vehicles in the DSRC are dynamic so that the attackers cannot track the identities of the vehicle. ITS-G5 uses ETSI TS 103 097 that is similar security standard to the IEEE 1609.2, which provides security to the vehicles.

8.2.3 ITS Spectrum Recommendation and Regulation Consideration

In recent years, a number of vehicle accidents and cyberattacks triggered a rapid awareness of security needs in the automotive industry. The EU, USA, and other countries developed several policies and initiatives on cybersecurity aiming to ensure the security of the smart and autonomous vehicles. More detail on the EU and international cybersecurity policy is given in Appendix C.

There is no doubt that 5.9 GHz band is the most appropriate frequency band used for ITS, considering the current development of the ITS ecosystem. In 3GPP, the frequency band 5.855-5.925 GHz is defined as Band 47 that is used for V2X applications. In 2015, the 3GPP started the study of C-V2X technology that consists of PC5 interface and Uu interface. The PC5 interface is a device-to-device communication over the direct channel, and Uu interface is the communication interface between the mobile device and the radio access networks. The C-V2X meets the demands of ITS applications such as rear-end collision warning, pre-crash detection, emergency vehicle warning, sudden stop, cooperative adaptive cruise control, road safety services, and automatic parking. The 5G in V2X offers sophisticated services such as platooning of vehicles, extended sensors, efficient driving, and distant driving. With the growth of C-V2X, more and more government agency and regulators show interest in C-V2X technology for future ITS developments. In regions or countries where technology neutrality has been incorporated, C-V2X tests or evaluations are recommended to check whether C-V2X complies with the relevant policy and regulations. If the regulations are compliant, the C-V2X devices in the allocated ITS spectrum should not be constrained by their service.

8.2.4 Cyber Security Standardization in V2X

The fast evolution and rise in the development of modern systems and networks, combined with the complexity of increasing risks, pose daunting challenges in preserving the security of vehicular systems and networks. The security solutions require a stable and safe network infrastructure; however, they also need to protect the privacy of the drivers as well as the vehicles. The cybersecurity is very important in applications such as autonomous vehicle because the life and property of the passenger are dependent on the autonomous vehicle security. It should be dynamic as the cyber-attack vectors are developing frequently and the autonomous vehicles should be protected from new cyberattacks. The security management needs proactive as well as reactive measures that are based on the dynamic approach.

The European Commission and the European Cyber Security Organisation (ECSO) engaged with the public–private partnership (cPPP) as part of the EU cyber security policy in July 2016. Its goal is to encourage cooperation at early phases in the research and innovation processes and to create cyber security solutions for different sectors, while establishing a European cybersecurity market and building

8.2 Standards, Regulations, and Legal Issues

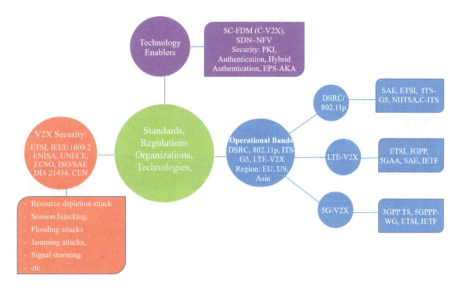

Fig. 8.4 Standards, regulations, technology enabled, operational bands, and security issues in V2X

the competitive industry in Europe. The PPP framework activities which is related to the security regulations affect the connected and autonomous vehicle technologies. The core activities are proposed for automotive vehicle domain to create a comprehensive cybersecurity risk management in vehicle automation application over the entire product life cycle. There are new areas of vehicle attacks due to the risks of cyber-attacks on vehicles, and weak online vehicle updates (OTA), V2X communication, and other requirements. The SAE is currently developing a new cybersecurity, i.e., ISO/SAE DIS 21434 "Road vehicles—Cybersecurity engineering" for cybersecurity of the vehicles, and the standard will be finalized in November 2020. The ISO/SAE 21434 aims to provide cybersecurity activities for all phases of the vehicle life cycle to manage cybersecurity activities and to analyze cybersecurity risk factors. The EU and US International Standards Harmonization Working Group established two Harmonization Task Groups (HTGs). They are Harmonization Task Groups 1 (HTG1) to harmonize security standards (including ISO, CEN, ETSI, and IEEE) to facilitate ITS cooperation and HTG3 to harmonize communication protocols. [11].

Figure 8.4 shows the overall existing standards, organizations, regulations, technology enablers, operational bands, and cyber security issues in V2X communication.

8.3 Competition Over V2X Technology Adoption

From time to time, there are some kinds of conflicts in V2X technology. In the past, the V2V technology was under severe threat from broadcast TV and other tech companies. They wanted to get a large portion of the radio spectrum from the bandwidth and use these frequency bands to provide high-speed Internet services, which were used by V2V.

In the area of smart transport, the FCC implemented the DSRC standard over two decades to ensure vehicle safety and performance on the roads. The EU and US federal government have proposed a mandate to implement V2V based on DSRC in new vehicles. In several regions, the DSRC using 802.11p standards was already in operation for traffic signals, intersection collision prevention, ambulance signal priority, toll collection, parking payments, etc. Soon, several countries will mandate to use V2X communications in new vehicles; however, there are no guidelines specifying how the event-critical messages should be transmitted. As the V2X technology advances, the V2X technology diverged into two different standards, DSRC and C-V2X, with fundamentally different architectures. This makes it difficult for original equipment manufacturers (OEMs) to choose which standard they should select and even more challenging is to harmonize a single global solution. The global automaker industry and academia are divided into two groups. One group is biased toward DSRC technology, and they think that C-V2X technology is not mature in vehicular communication field compared to DSRC. It might take some time for C-V2X to be widely deployed, as it has not been extensively tested in the field. The other group supports C-V2X technology due to its successful implementation in mobile networks, its potential, and evolution toward 5G.

On the one hand, we have matured and successfully tested DSRC for V2X communication which has no room for evolution, while on the other hand, we have a new and promising C-V2X technology which has potential to evolve beyond 5G. Although C-V2X was introduced recently as compared to DSRC, it provides better performances in terms of communication range, latency, reliability, etc [12]. The C-V2X offers QoS benefits besides worthy performances in carried-out tests and trials [13]. In addition, C-V2X can coexist with other networks or applications in a 5.9 GHz band on the co-channel and/or an adjacent channel.

It would be better if we can use both the technologies for V2X communication. However, if they run on the same frequency band, there is conflict with each other because there are differences between wireless systems. Basic differences between DSRC and C-V2X are given in Table 8.3 [14].

The decision to adopt either DSRC or C-V2X is split around the globe. The automotive industries such as Volkswagen, Toyota, General Motors, NXP, and Volvo are supporting DSRC technology for V2X, while the 5GAA and other industry and institutes like BMW, Ford, Daimler, Qualcomm, Intel, Samsung, and Huawei back up 5G and support for C-V2X adoption because of its advantages and evolution in V2X communication. In Europe, there has been a heated debate on the adoption of DSRC over cellular technology. In July 2019, the new European Council (EC) of minister advocates 5G over DSRC reversing the previous EU commission decision on April 2019. As of writing this chapter, the newly elected EC will have to draft a novel plan with a "neutral" approach to the Council, enabling car manufacturers and operators to settle on what standard to use for connected cars. Even though the decision is positive toward the 5G industry, the battle for adoption is not over yet.

8.3 Competition Over V2X Technology Adoption

Table 8.3 Basic differences between DSRC and C-V2X

Components	DSRC V2X	Cellular V2X
Technology	Wi-Fi	LTE/5G
Communication technology	OFDM with CSMA provides robust communication in dense and dynamic environment. No dependency with GPS signal.	SC-FDM with semi-persistent sensing, self-managed with ability for centralized management
Cellular connectivity	Hybrid model, i.e., connect with cellular network for non-safety services	Hybrid, cellular network
Transmission scheduling	CSMA: No predetermined Tx slots and transmit when there is no ongoing reception	Collisions are not sensed. Slow response to changing environment
Line coding	Convolution code	Turbo code
Deployment	From 2017. Commercialization in 2019	Mass market deployment in China from late 2020
Future guideline	Backward compatible and interoperable upgrade from 802.11p to 802.11bd	C-V2X Rel.16 is based on 5G NR technology and operates in different channels than previous releases
Latency	Low latency for V2V communication	Round trip latency less than 1 ms, slight delay due to centralized communication
Range	Good for short radio range	Optimized for long-range communication

According to EC transport commissioner, DSRC is already a proven and secure technology that can be implemented easily, cheaply, and rapidly ensuring that the large amount of already invested resources are not waste. If implemented earlier, the technology can improve the road safety and save the lives of people on the road before facing further severe accidents. It will take more than three years for the new technology like 5G V2X to be ready and deployed. It is necessary to install 5G supporting chipset and beacons on road signs, traffic lights, RSU, emergency vehicles, etc., for V2X services. There has not been sufficient rigorous testing for 5G-based V2X as compared to DSRC. Others think that the V2X communication should be free. If 5G technology is used as a platform in V2X, then who will pay for the 5G SIM card, and supplementary revenue streams for using 5G need to be paid to the cellular operators. The basic concept of V2X is that consumers use the technology for safety features and cellular operators should not hold them as hostages of using the 5G technology.

Hence, the V2X standards require a global solution. The V2X has the ability to perceive its environment accurately, i.e., situational awareness even when autonomous vehicles do not have visibility and other complementary sensors. A global V2X solution will open the way for autonomous vehicles anywhere and will improve their widespread acceptance.

Autotalks, an Israeli-based provider of V2X Chipmaker Company, already launched a global V2X solution that incorporates a cellular V2X feature with intrinsic support for DSRC. They manufactured the world's first dual-mode C-V2X chipset that supports both cellular and DSRC technology [14]. C-V2X and DSRC are equipped with one chip for the OEMs. They mentioned some of the advantages of OEMs are low development costs, feasibility of certification, global single certification, alignment, and cost effective between V2X and mobile networks. It is also beneficial to the consumers in terms of safety and security, ensuring the use of safety and infotainment messages and pay-as-you-go scheme while using C-V2X. In June 2019, Autotalks and China's Datang, a leader in developing C-V2X standards and solutions, successfully tested the interoperability of its respective C-V2X direct communication solutions at the chipset level for C-V2X direct communication. The test validates both companies' solutions to inter-work according to 3GPP Rel. 14 standards, which is a requirement for mass deployment of C-V2X technology. The testing is a milestone and shows the competency of the Autotalks and Datang stand-alone PC5 solutions [15].

Whatever technology is used for V2X communication, whether it is DSRC or 5G or both, the security features need to be ready. Recently, Autotalks announced in their Web site that their second-generation chipsets are Federal Information Processing Standard (FIPS) certified for secure C-V2X/DSRC deployment in the USA. The information processing is performed thinking about the autoregressive moving average models.

8.3.1 Challenges for DSRC V2X and Cellular V2X

In this section, we will briefly discuss the challenges faced by cellular and DSRC V2X technology focusing on latency, security, privacy, capacity, etc. There are still many issues to address when collaborating with different business partners to develop new V2X technologies. The "knowledge gap" between network operators, academia, and vertical businesses could be the critical obstacle. While network operators have extensive knowledge of 5G wireless technology, they have not obtained technical knowledge in each academic and industry. Vertical companies are well aware of the need for technical transformation, although the actual implementation scenarios for V2X technology remain difficult to deal with. We present the challenges faced by cellular and DSRC V2X technology in brief.

The DSRC based on 802.11p uses a short-range peer-to-peer communication, which is good for connecting with neighboring vehicles in ad hoc modes. However, the short-range communication and lack of pre-existing DSRC-installed RSU result in inaccessible connectivity due to limited network coverage in sparse network scenario. The line of sight (LoS) and short-range communication have consequences for effective coverage. The LoS conditions are a concern for the implementation of V2X communication, particularly in urban settings where major LoS obstacles such as buildings and trees are the norm instead of the exception. Under normal operating

8.3 Competition Over V2X Technology Adoption

conditions, latency is not a particular concern for 802.11p. In suboptimal conditions, an increased packet error rate and the corresponding requirement to retransmit messages may lead to a higher latency. The vehicular traffic congestion can lead to a high channel bottleneck easily and have a major impact on packet error. Small range of data rates between 6 and 27 Mbps is very small and could not support potential future V2X applications. Improved congestion control systems and congestion control schemes can provide a potential path to solving the congestion problem. The 802.11p-based DSRC is vulnerable to several types of attacks because of the ad hoc characteristics. The potential attacks on DSRC are authenticity, availability, confidentiality, and integrity. Some of the security issues can be mitigated by VPKI deployment and distributed misbehavior detection; however, many potential threats, such as vehicular worms and wormhole attacks, are difficult to prevent. The use of short-term pseudonymous certificates to authenticate V2V communications provides privacy to DSRC nodes. Advanced eavesdropping and data collection may still pose a risk to vehicle passenger privacy. The absence of existing DSRC infrastructures on the road and the need for additional DSRC module in every vehicle incurs substantial economic cost for both authorities and vehicle users.

In case of cellular V2X, during the initial deployment, the 5G coverage cannot be penetrated in the rural villages or mountainous locations; as a result, the C-V2X communication becomes inconsistent. However, using sideling D2D communication can be a potential solution for V2V coverage. The latency in C-V2X increases due to the processing delay through the infrastructure nodes such as eNB and Evolved Packet Core (EPC). Some of the solutions to these problems are to adopt D2D side link and local edge resources to reduce the latency. One of the challenges the C-V2X has to suffer is significant message transmission congestion due to the frequent unicast transmission through the eNB nodes from several hundreds of vehicles. The 5G aims to support very high data rates that might be adequate for V2X applications, and the use of eMBMS or side link D2D might solve the congestion issue. As for security, the long existence of cellular network for mobile communication shows strong resistance to vulnerable attacks. However, it is not sure if the same security techniques can be directly used in C-V2X communication, as some attacks such as jamming and DoS attack are difficult to overcome. One of the privacy issues with the C-V2X is that the network operator or the authority in cellular communication stores the subscriber ID and all other information of the user. If the network operators behave maliciously or if it is under attack, then the user privacy will be compromised. The mobile network infrastructure is already in service with a broad coverage range, high capacity, and high throughput, but there is financial obstruction to support seamless V2X nationwide coverage through the cellular networks. The installation of cellular radios in all the vehicles increases the additional economic cost for the vehicle users. In addition, some cellular network operators charge a significant rate for data usage while the DSRC-based V2X communication is free of charge. Table 8.4 shows the challenges in DSRC V2X and C-V2X technology.

Table 8.4 Challenges in DSRC V2X and C-V2X technology

Features	DSRC V2X	C-V2X
Coverage	Short-range and limited network coverage in sparse network	Limited coverage and inconsistent communication in mountainous locations
Latency	Packet error rate and retransmission increase latency	Processing through infrastructure (eNB, EPC) increases latency
Capacity	Small data rates between 6 and 27 Mbps may be insufficient for future V2X application	Frequent unicast transmission through eNB from hundreds of vehicles causes congestion
Security	Several threats like vehicular worms and wormhole attacks are difficult to prevent	Some attacks like jamming in CV2X are still difficult to tackle
Privacy	Malicious data collection and advanced eavesdropping could pose a risk to the privacy of vehicles.	Leakage of subscriber information from network operator causes privacy issue
Cost	Installation of DSRC-based infrastructures and modules on each vehicles incur substantial economic cost	High data rate and installation of cellular radios in each vehicles cause higher cost for vehicle users

8.4 V2X Use Cases

8.4.1 Smart Mobility

In the era of the Fourth Industrial Revolution, the smart mobility will strive for sustainable mobility and new means of transport such as shared autonomous vehicle and shared electric vehicles. Smart mobility is an integral concept for future intelligent transportation system that has become smarter and more intelligent with the integration of current transportation system and advanced functions of smart devices. The auto industry is attracted toward the smart mobility ecosystem allowing specific implementation of those solutions. The smart mobility should be designed wisely considering the intelligent and autonomous vehicles. In smart mobility, the vehicles gather on the fly to be part of a social network for smart mobility, i.e., a network of neighboring vehicles. The vehicles that share the same preferences, places, and travel in the same direction or destination become part of existing networks. There are intelligent tools available that use real-time traffic statistics, obtained through collaborative sensors, to help vehicles select the best path. Some ridesharing apps such as Uber and Lyft enable vehicle owners to share their cars with those using similar destinations, and this can be viewed as the concepts of smart mobility.

8.4.1.1 Smart Mobility and Security Issues

In this section, we will discuss different types of smart mobility and its associated risk factors and security issues. There are four states of mobility in smart mobility. The

8.4 V2X Use Cases

Fig. 8.5 Smart mobility states

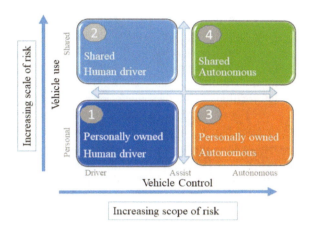

states represent current and future autonomous driving as well as vehicles/ridesharing platforms as shown in Fig. 8.5. The four states of smart mobility present an innovative way to enhance the mobility features in intelligent and autonomous vehicles [16].

1. The first state represents the current status of the personally owned vehicles where the human driver operates the vehicles. The most advanced vehicles in this state are able to connect with other vehicles and avail features of advance driver assist technology. The vehicles in this state are increasingly becoming data centric; i.e., they can create, consume, share, and analyze the information. There is cyber security risk related to the advancement of the vehicles in this category. The vehicles can be hacked by the attacker by intercepting wireless communication and tracking its location using various attack tools, then take over the access control of the system, and then might rob the important equipment or accessories of the vehicles. Due to this, the attack impact and the financial damage are much larger. Some of the advanced vehicles already employed a certain level of enhanced security features based on in-vehicle technology.

2. The second state represents the shared vehicles that is controlled by human driver. Along with the growth of ridesharing companies that boost the sharing economy, such as carsharing and kickboard sharing, the shared mobility industry continued to expand. The ridesharing company such as Uber and Lyft and carsharing company such as Zipcar and Socar provide a glimpse of cleaner, greener, cheaper, safer, and convenient future transportation system. The ridesharing and carsharing can be done through the smartphone applications where the users provide their credit information and bank information for payment. However, the exponential rise of ridesharing without considering a strong security shield, will dramatically increase the magnitude and complexity of cyber threats. With vehicular social network (VSN), all the vehicles are capable of using social media, different types of mobile applications as well as ridesharing apps, and this leads to additional cyber security risks. The hackers can compromise the onboard communication system and obtain location information; as a result, they get information

regarding the payment system such as credit card information and banking information including other personal information. If the private information such as bank information or information related to health and insurance is revealed to the attacker, then they might misuse the information like transfer all the money from the user's account to the attacker's account. Hence, securing the personal information of both the vehicle and the driver becomes a very high priority.

3. The third state represents the future state of personally owned vehicles, but without human drivers, i.e., fully autonomous vehicles. The future vehicles will be fully functional autonomous vehicles based on different types of inbuilt sensors and electronic control units (ECUs) that communicate with each other. The passenger's experience will extremely improve with autonomous features of the vehicles. The personally owned vehicles are comparatively less susceptible to hackers if strong security measures are taken. The autonomous vehicles are also required to communicate with other vehicles and infrastructures based on V2X communication and other external systems, but they are likely to open new vulnerabilities. In personally owned fully autonomous vehicles, the attacker can use wide range of attacks from simple attacks, where malicious messages are sent to a vehicle, to more sophisticated attacks in which attackers may open up the ECUs and try to reverse engineer their microcontrollers and software. If the attacker takes control of the vehicle system such as steering and braking systems, then the damage caused by the attacker would be very critical and might cost the life of the passengers.

4. The fourth state represents the future state of shared vehicles without the drivers or fully shared autonomous vehicles. In the coming future, there will be advanced ridesharing technology, where the individual riders simply request the rides from the nearby autonomous vehicles. The vehicles will take the passengers to their destinations easily and without any hassle. The future autonomous vehicle ridesharing could bring the revolution in the smart mobility considering the smart city. The autonomous ridesharing will motivate and bring great opportunity to integrate different technologies and mobility ecosystem that offers cleaner, greener, and advanced form of transportation system. Similar to the second state, there are security and privacy issues related to the ridesharing of autonomous vehicles but higher order of magnitude. The attackers might target the critical objects, spoof the GPS information, and provide false location information as well as cause jamming to the sensors or DoS attacks that disrupt the vehicle sensors from getting critical information. The attackers can target different sensors such as LiDAR, RADAR, camera, ECU, GPS, and other in-vehicle and inter-vehicle communication system that can disrupt the proper functioning of the shared fully autonomous vehicle. The autonomous vehicles and ridesharing system should be more secure against various types of cybersecurity attacks that enables the future of smart mobility more secure and resilient.

8.4.2 V2X Testbed

In 2013, the 5G forum was founded in South Korea. South Korea became the first country to commercialize world's first 5G communication networks. In 2019, a collaboration with 5G forum and ITS Korea was held jointly by 5GAA. The 5G forum is a public–private forum, which promotes 5G use cases, 5G commercialization, and 5G convergence services. The 5G forum intends to be the leader of next-generation communications technology and promotes the economic growth through the development of the ICT sector in attempts to update the creative economy's new administration agenda. In the same year, the Seoul Metropolitan Government collaborated with SK telecom constructed the world's first and only testbed as shown in Fig. 8.6. The testbed is based on 5G and V2X communication for urban autonomous vehicle driving innovations in the area of public transportation. The Seoul city created the K-City as a smart city testbed, a Digital Media Street (DMS) in Seoul. However, we will not discuss the smart city and IoT in this chapter, as it is a wide topic. In this chapter, we will only focus on the 5G testbed for C-V2X communication.

The testbed includes a CCTV control management device that can track and measure both autonomous driving conditions and situations. It consists of 5G and V2X communication networks, high-definition (HD) 3D maps, vehicle maintenance, parking lots, charging stations for electric cars, future mobility center, and all the other services and conveniences required for testing the autonomous driving technology. The testbed is established for autonomous vehicle driving and pilot operation. The SK Telecom successfully demonstrated the autonomous 5G V2X convergence driving technology use case and provided a model of 5G-based V2X technology that makes it possible to operate safely, even under adverse weather conditions,

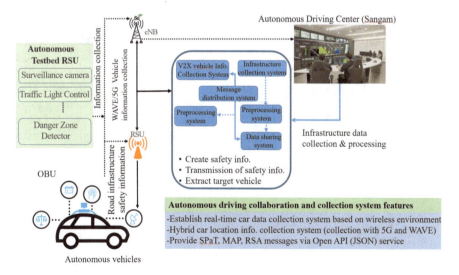

Fig. 8.6 G V2X tested for autonomous vehicles

which could otherwise be difficult to drive completely only depending on sensors. In the testbed, the 5G advanced driver-assistance Systems (ADAS) are installed in public vehicles to support V2X communication that communicates with RSU and other adjacent vehicles in V2I and V2V mode [17]. The high-speed 5G network becomes mandatory as the information generated by the number of connected vehicles increases exponentially. The 5G V2X communication testbed and control center for autonomous vehicle used in Seoul DMS are given in Fig. 8.6. The autonomous driving center is responsible for collecting and processing all the information from the autonomous vehicles and the autonomous testbed road infrastructures. It established real-time vehicle data collection system based on wireless communication environment using 5G and WAVE communication technology. It provides SPaT, MAP, RSA messages through open API services. The HD 3D maps are used for navigation and location information, which is possible through 5G V2X communication. The driving center helps in creating safety information system and transmission of the safety messages to the autonomous vehicles in real time. The 5G V2X testbed facilitates robust research and development for autonomous vehicles as well as real-world performance evaluation and international accreditation testing.

Similarly, in other countries, the race for 5G C-V2X testbed is also developing. The European Commission under the Horizon 2020 founded 5G HarmoniseD Research and TrIals for serVice Evolution (5G DRIVE) research and innovation action between EU and China that runs from 2018 to 2021. The 5G-DRIVE comprises 17 European partners from 11 countries and from various sectors. The trial test plans offered by the two V2X experiment-testing sites, Espoo in Finland and Ispra/JRC in Italy, combine two main test types. At the Espoo trial sites, the vehicles must drive in a real environment, thus requiring more real testing situations for evaluating 5G advantages, especially with URLLC. At Ispra, trial site is presented in a setting that is controlled based on harmonized standards. This combination aims to provide the necessary variation of tests to evaluate the eligibility and advantage of 5G for V2X scenarios. In Singapore, Nanyan University of Technology (NTU) and Local Telecommunications Company are working together to incorporate 5G technology and V2X communication testbed, which is first of its kind in the country [18]. The latest testbed will also enable industry partners to cooperatively design and deploy 5G connectivity for a variety of safety-critical applications such as collision prevention, real-time traffic routing, and network security. The C-V2X devices will install in the buses and in autonomous vehicles that are already in operation on NTU Smart Campus site.

8.5 Current Trends and Future of Intelligent and Autonomous Vehicles

We expect that this chapter will contribute to existing and potential debates and lead to develop forthcoming approaches of future intelligent and autonomous vehicles. As soon as the technical and regulatory matters are resolved, more than 15% of new cars

8.5 Current Trends and Future of Intelligent and Autonomous Vehicles

sold in 2030 might be fully autonomous vehicles [19]. The future intelligent and automotive vehicles will be electrified, fully autonomous, shared, and fully connected. It will emit very less smoke and less noise because it is electric and consumes less time and space because it travels autonomously, it will be economical or affordable as people will use shared autonomous vehicles and everybody will use it, as users will not need a driving license to use it. The behavior of consumer mobility is evolving, and there is a sudden rise of automotive industry solution that gives rise to one out of ten vehicles sold in 2030 being possibly a shared vehicle. Motivated by shared mobility, communication services, and functionality updates, new technologies and business models could increase the pools of automotive revenue by about 30%.

8.5.1 Trends in Intelligent and Autonomous Vehicles

The four trends play a vital role in the development of future autonomous vehicles. The key four trends are autonomous driving, connectivity, diverse mobility, and electrification of the vehicles as shown in Fig. 8.7. The riders and drivers use vehicle connectivity and subsequent autonomous technological service models to become a medium for their transport and for personal purpose. The accelerating progress allows automobiles to be upgradable particularly in software-based systems. When shared mobility services such as ridesharing or e-hailing, with shorter development cycles, become more popular, the customers will be constantly aware of advances in technology, which will further increase demand for upgradability of privately owned vehicles. In addition, strict emission standards, lower battery prices, commonly available

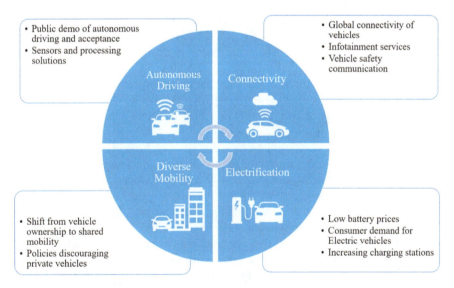

Fig. 8.7 Basic trends of future autonomous vehicles

charging points, and increased consumer awareness will develop new and growing support for the adoption of electric powered vehicles such as hybrid, plug-in, battery electric, and fuel cell, in the coming years, which will vary significantly at regional and local levels.

1. Autonomous Driving: The exponential progress that has been made in fields like artificial intelligence, machine learning, and deep neural networks makes it easier to accomplish autonomous vehicles that do not need human interference except in complex traffic scenarios. The use of human accessibility systems should be fully reshaped. New features such as 360° sensing using various sensors, and human like driving behavior using high-computing onboard system and advance cybersecurity system are emerging, which would have been unimaginable just a few years ago.
2. Connectivity: The connectivity services in vehicles provide networking of vehicles with other vehicles and infrastructures such as RSU and clouds as well as provide Internet connectivity for Web surfing, work, or accessing multimedia services during the journey. The vehicles use connectivity service for getting critical information updates so as to prevent from serious accidents as well as it can be used for infotainment purposes. There are several cybersecurity issues and advance form of techniques used to combat the security issues, which are discussed in detail in Chap. 6.
3. Diverse Mobility: The shared mobility has been offered by many developed countries. However, sharing concepts of autonomous vehicles will be economically viable soon. The quest for a shared vehicle in the local area would no longer be necessary: Rather, vehicles will be requested anywhere the consumer happens to be through a convenient "on-demand" service know as mobility-as-a-service. Moreover, the product set will be revised periodically to reflect the newest advances in hardware and software as the basic vehicle chassis will be the same. Because of the high cost, consumers would obviously not choose to buy a new vehicle each year; the short innovation cycles would primarily reach the market through frequent updates of shared vehicles. Autonomous electric taxi or driverless robotaxi might be available in the future for transportation facilities.
4. Electrification: The shift to emission-free autonomous vehicle would not be possible without electrification of the vehicles. There is problem even with a very small amount of harmful substances, dust, and noise generated by the current vehicles operated on non-renewable gasoline fuels. Although it will take some time to replace huge percentage of vehicles that operate on fuels that was introduced more than hundred years before, the development of electric vehicles is evolving in an exponential way. This also suggests that "emission-free" electric vehicles should use green and renewable source of electricity to charge the autonomous vehicles. Because of the exponential development of electrical-driven vehicle technologies, the vast majority of automated vehicles beyond Level 4 can be considered autonomous e-vehicles. The autonomous and sharing vehicles would make a significant use case of electrified vehicles in the future. Several automotive companies like BMW, Honda, Uber, Tesla, etc., are planning

to develop and implement robotaxi, which is a driverless taxi operating on or over Level 4 for on-demand mobility services soon.

Ericsson, one of the members of 5GAA, addressed four automotive megatrends called Connected, Automated, Redefine and Electric (CARE) that will determine the future of intelligent and autonomous vehicles. The concept is similar to the one defined in Fig. 8.7. The connected vehicles exist at the center of the transportation system, while the automated self-driven vehicles assist in the transportation of the people in the smart city. The current business model should be redefined to match with the ever-evolving business models like shifts from personalized vehicles to sharing of vehicles. In addition, electrification of all the future vehicle-supporting cleaner and greener transportation system will ultimately lead to safer and cleaner smart cities.

8.5.2 Autonomous Electric Vehicle and Challenges

In this section, we will discuss about the future autonomous electric vehicles and its challenges. As electric vehicle (EV) topic in itself is very wide, we will discuss only the general concept and future trend.

Several countries in the world propose to a ban on the selling of gasoline and diesel vehicles by shifting to electric and hybrid vehicles. Countries in Europe like Norway prohibit fossil fuel vehicles and create opportunities for electric vehicles. All other countries will also follow this trend for cleaner, greener, and healthier environment, while fossil fuels have limited supply. The future will be full of autonomous and electric vehicles over time, and it will likely become the preferred form of transportation with the goal of moving ahead with more sustainable and efficient e-mobility alternatives. The charging facilities would be a critical factor in maintaining a seamless transition to e-mobility. Innovations with novel EV charging solutions are required that would play a key role for EVs. There have been a few ideas that are going around over the years with a few that appear to stand out such as smart charging, contactless charging, vehicle-to-grid (V2G), on-road EVs charging, and photovoltaic (PV) panel for EV charging. Cities are already trying to enforce regulations, which require new homes and even new parking lots to have electric car charging facilities. Some automakers such as Volkswagen proposed a robot that carry small portable charger to the designated parking spaces, and they move only when the requested user calls them. Similarly, Tesla Inc. has installed more than 16 K supercharger all around the globe and is still increasing to ease the autonomous electric vehicle in the future.

However, there are several challenges in autonomous EVs. The key obstacles for electric vehicles are limited travel range, high prices, battery failures, and a spotty charging system. Furthermore, there are concerns with specific power semiconductors and other devices. There are safety and security concerns with EVs as EVs are run on the battery, and there might be charging leakage that may cause electric shocks or the battery might explode due to issues in the battery manufacturing. Another technical challenge is lithium-ion batteries because the lithium-ion technology is reaching its limit, other battery technologies are only in research and development, and they are not expected sooner. Moreover, the elements required for manufacturing battery such as lithium and cobalt supplies need to increase to meet the demands of ever-rising EVs. Currently, the number of stand-alone EV charging stations is limited and needs to scale up to recharge hundreds of millions of EVs at the same time in the future. In addition, EVs consume lots of charging time to get fully charged and on top of that, the EVs cannot travel very long distance on a single charge. As all the automotive industry will manufacture EVs, they must comply with same type of compatible charging ports and charging stations. It is also important to improving the performance, durability, and cost of subsystems and equipment.

8.5.3 Cyber-Attacks in Future Autonomous Vehicles

Currently, not all the people are ready to accept and ride the autonomous vehicle due to lack of full trust on the autonomous vehicles. The autonomous vehicle should first gain the human trust by performing excellent test drive better than human driver without any accidents on the street. The vehicle manufacturers have a strong record of accomplishment in setting up a vehicle safety, but not in cybersecurity so far. People still doubt on the seamless vehicle connectivity, privacy, and data security including the security of biometrics information produced and shared by the connected vehicles. Systematic and stepwise threat modeling methods such as STRIDE and PASTA need to be used to prevent the autonomous vehicles from cyberattacks, and more details on this can be found in Chap. 3.

8.5.3.1 Protection of Future Autonomous Vehicle from Cyber Attacks

The vehicle manufacturers must be encouraged to include cybersecurity in design from the beginning and over the complete vehicle life cycle as the new cybersecurity risks can arise at any time. Anything that happened in our car usually stays in our car in the past; however, that is not the case anymore. The rapid development of infotainment, over-the-air (OTA) software updates, and other technological developments are converting vehicles into technology warehouse. The technological transformation of the automotive industry driven by new vehicle connectivity, individual mobility concepts, autonomous driving, and electric vehicle poses new cybersecurity challenges as well. The hackers and other black-hat attackers attempt to gain access to critical computer systems and data in vehicles that is likely to jeopardize the safety mechanisms and user privacy. Some of the cyber threats in vehicle services are vehicle apps, online services, e-vehicle charging stations, and online vehicle features that connect to onboard vehicle system remotely. Many users would need a new degree of trust in security, safety, integrity, and reliability of autonomous vehicle technology and its infrastructure to ensure that their safety is in the hands of automated vehicles.

8.5 Current Trends and Future of Intelligent and Autonomous Vehicles

It is anticipated that autonomous vehicles can have multiple onboard attack vectors including radar, cameras, GPS, ultrasonic sensors, V2X, infrastructure devices, and applications that these sensors rely on. The technical and functional criteria for fully automated vehicles are considerably higher than those for partially automated or assisted driving vehicles. When everything is connected, it can also interrupt everything, with cyber attackers continuously searching for advanced illegal techniques to use the Internet for their personal financial gain.

In the future, the autonomous vehicles will have approximately 300 million lines of software codes. The explosion of complex software code gives cyber attackers to attack not only in the vehicle, but also in the entire automotive industry [20]. Currently, there is no standard approach to fight against the cybersecurity in the vehicle industry. The development of electrical, electronic, and digital network infrastructure in vehicle increases surface attacks and tends to increase the cyber risks. Some of the emerging cyberattacks in vehicle system are given below:

1. Spoofing of sensors: The attacker can attack the vulnerable sensors and access the autonomous functions of the vehicle to control the engine and braking systems. This might result in serious accidents.
2. Hijacking: The attacker can hijack and take control over the safety control system like braking systems and engine characteristics.
3. Surveillance attack: The attacker can hack the voice recognition systems such as mic and listen to the voices of the drivers/passengers in the vehicle, or they might take over the vehicle black box system.
4. Physical access: The hackers manipulate the security system and gain secure access directly though the OBU to manipulate the car information and tuning the integrated chips.
5. Infotainment system: The attacker gains control of the external wireless connectivity system such as Wi-Fi, Bluetooth and NFC and accesses the infotainment systems.
6. Telematics: The attacker remotely unlocks the doors of the vehicles by manipulating the external connection system.
7. DoS attack: The attacker can cause DoS attack on the back-end service devices like servers or cloud connectivity and stop the functionality of the vehicles.
8. OTA: The attacker attacks on the weak online vehicle software updates based on OTA and misuses or compromises the update procedures, or denies the legitimate updates. I t ncreases the risks of cyber-attacks on vehicles.
9. Unauthorized access: The attacker attacks the weak OEM back-end services, gets unauthorized access of the user management function, and manipulates the user data.
10. Information stealing: The malicious attacker gets access of the unsecure vehicle third-party front-end devices like music player, mobile apps, etc., through the devices connected to external interfaces like USB ports or OBD ports. They obtain vehicle owner's private data through these devices.

8.5.3.2 Cybersecurity of Future Vehicles and Machine Learning Techniques

Machine learning (ML) has proved helpful in combating cyber-physical attacks that endanger vehicle network safety and security. The current and future semi-autonomous and fully autonomous vehicles will be heavily dependent on artificial intelligence (AI) systems. Deep learning with convolutional neural network (CNN) works best for the identification of path, line, and obstacle. It is used to detect boundary boxes, and the result is instantly passed on to the other algorithms, which determine whether the vehicle is the same as the previous one. Some of the ML techniques that can be used in intelligent and autonomous vehicles are discussed in Chap. 3.

The next generation of AI that will be based on human–machine interaction, such as artificial superintelligence (ASI), uses data and advanced algorithms to imitate the human brain cognitive functions. They can learn, understand, and resolve issues independently. The autonomous vehicles based on ASI systems imitate, augment, and facilitate human behavior, while simultaneously exploiting rapid response times and precise machine-controlled technology. The intelligent vehicle also consists of several types of sensors along with LiDAR and camera sensors that can detect 3D images of people on the road, neighbor vehicles, road signs, animals, and buildings and utilize high-definition 3D maps. In the future, various types of advanced ML, ASI, reinforcement learning, and computer vision technology can be used together to control specific sensors and collaborate via advanced algorithms. This can provide robust solutions and imitate human like logical thinking for perception problems in real-world environment.

8.5.4 Challenges in Future Autonomous Vehicles

The machine learning also poses potential challenges and risks by itself, as the system based on ML can generate adverse or unpredictable results and can cause accidents. For example, convolution neural networks can easily be manipulated by maliciously constructed noise images. On the other hand, reinforcement learning agents seek undesirable ways of enhancing the reward function provided by their interacting environment. In such conditions, if the accidents are caused by the autonomous vehicles, then who is guilt and takes responsibility of the accidents, the autonomous vehicle itself, the driver behind the control system, or the vehicle manufacturing company. There might be situation where the autonomous vehicle has to divert from the usual path to save six people on the crossroad and might kill one on the other side of the road. It creates a new ethical challenge. Such unfortunate situation brings the dilemmas that make it difficult for people to accept that ethical dilemmas occur mostly in severe and disastrous circumstances. New rules and laws should be drafted, and legal actions should be taken considering the advancement of autonomous vehicles because introduction of autonomous vehicles might reduce more severe accidents and disasters on the road in the future.

The autonomous vehicle cybersecurity is still a new area, and the cybersecurity threat on the autonomous vehicle will remain growing issues. One of the challenges is to design a robust cyber security framework for core 5G autonomous vehicles to protect the critical information and communication infrastructure. The vehicle manufacturers need to consider the vehicle cybersecurity as an essential part of operations and development efforts. The autonomous vehicle industry needs to adopt OEM cybersecurity issues, build skills to implement cybersecurity best practices into its products, and work effectively with OEMs to integrate and validate end-to-end cybersecurity solutions. Consequently, the autonomous industry must create a common cybersecurity standard that provides end-to-end cybersecurity solutions that keeps the expenses of development and maintenance under control.

8.5.5 Intelligent Autonomous Vehicle Improves Environment

In the USA, ITS transportation sector had contributed 27% of the total greenhouse emissions, in 2013. The electrified intelligent autonomous vehicle with V2X technology will allow drivers to make better decisions while driving; for example, information about traffic status will help drivers to take better routes and avoid unnecessary stops, overall improving the fuel efficiency of the vehicles. The vehicle drivers maximize their plans by getting information about public transportation in real time. Automated eco-driving techniques will also be helpful in reducing environment degradation. Eco-driving is a set of practices that aims at reducing the use of fuel consumption in the vehicle without changing the mechanics of the vehicle. This can be achieved by optimizing the speed and the amount of cylinders used in order to maintain adequate power output, and by optimizing the acceleration and deceleration behavior of the drivers. It has been observed that quite a few human drivers are using incorrect acceleration or deceleration techniques. Either they accelerate too quickly or they brake too forcefully. It results in the electricity and fuel consumption being wasted. If the vehicles use these eco-driving methods, the energy consumption of each vehicle on the road is projected to be decreased by as much as 10 to 20%.

8.6 Summary

In this chapter, we discussed the V2X current security issues, standards, challenges in Europe, USA, and other countries. We first described several different types of standardization, associations, and organization working in DSRC and C-ITS protocols. Then, we introduced V2X technology based on DSRC and cellular network and its adoption. We also discussed different types of smart mobility and its associated risk factors and security issues as well as 5G V2X testbed and its use cases.

It addresses the effect of electric vehicles in intelligent and autonomous vehicle and carsharing applications. This chapter introduced carhailing and ridesharing services as a promising solution to minimizing private vehicle utilization in a community, thus minimizing the need for parking spaces, reducing traffic congestion, and contributing to emission reductions. Lastly, we presented the future of intelligent and autonomous vehicles, electric vehicles, and its cybersecurity issues.

References

1. I. Jemaa, P. Cincilla, A. Kaiser, and B. Lonc, "An Overview of Security Ongoing Work in Cooperative ITS," in *12th ITS European Congress*, 2017
2. Car2Car, "Car 2 Car Communication Consortium." [Online]. Available: https://www.car-2-car.org/
3. SAE, "Standards Collections," 2020. [Online]. Available: https://www.sae.org/standards/. [Accessed: 15-Jan-2020]
4. UN Task Force, "Draft Recommendation on Cyber Security of the Task Force on Cyber Security and Over-the-air issues of UNECE WP.29 GRVA," 2018
5. GAA, "5GAA: paving the way towards 5G." [Online]. Available: https://5gaa.org/5g-technology/paving-the-way/. [Accessed: 17-Feb-2020]
6. G Automotive Association, "The Case for Cellular V2X for Safety and Cooperative Driving," *5GAA Whitepaper*, pp. 1–8, 2016
7. D. P. F. Möller and R. E. Haas, *Guide to Automotive Connectivity and Cybersecurity*. 2019
8. ACEA, "Frequency bands for V2X," 2018
9. TechVision Group of Frost & Sullivan, "Vehicle-to-Everything Technologies for Connected Cars: DSRC and Cellular Technologies Drive Opportunities.," 2017
10. GAA, "Timeline for deployment of C-V2X – Update," 2019
11. EU-USDoT, "EU-US standards harmonization task group report : feedback to standards development organizations - security," New Jersey Avenue, SE Washington, DC, 2012
12. NGMN, "Liaison Statement on Technology Evaluation of LTE-V2X and DSRC," 2017
13. GPP TR 22.886, "Study on enhancement of 3GPP Support for 5G V2X Services," 2016
14. Autotalks, "One global V2X solution: DSRC and C-V2X," *Autotalks*, 2019. [Online]. Available: https://www.auto-talks.com/technology/global_v2x_dsrc-and-c-v2x/. [Accessed: 22-Aug-2019]
15. Autotalks Ltd., "Autotalks and Datang Successfully Completed Chinese C-V2X Direct Communication Interoperability Testing," *Press Releases*, 2019. [Online]. Available: https://www.auto-talks.com/3434/. [Accessed: 25-Jan-2020]
16. "Securing the future of mobility Addressing cyber risk in self-driving cars and beyond."
17. J.-H. Jun, "5G- Powered traffic system," *Korea Times*, Seoul, Korea, pp. 1–1, 17-Jan-2019
18. NTUS, "NTU Singapore to develop vehicular communications for multi-modal mobility solutions," *Press Releases*, 2018. [Online]. Available: http://news.ntu.edu.sg/pages/newsdetail.aspx?URL=http://news.ntu.edu.sg/news/Pages/NR2018_Oct181018-3158.aspx&Guid=f760f09d-0189-4203-828f-aacd83c25449&Category=News+Releases
19. McKinsey&Company, "Advanced Industries: Connected car, automotive value chain unbound," p. 50, 2014
20. J. Deichmann, B. Klein, G. Scherf, and R. Stützle, "The race for cybersecurity: Protecting the connected car in the era of new regulation," no. October, 2019

Appendix

A. *International cybersecurity standard effort in automotive industry in Asia and other parts of the world*

In Japan, the Society of Automotive Engineers of Japan (JASO) has introduced "JASO TP-15002: Guide to Analyzing Vehicle Information Security" in 2013 and continues. Its main goal is security analysis methods such as threat analysis and risk assessment for vehicular systems. In addition, the Japanese Ministry of Economy, Trade and Industry is taking initiatives and studying measures for cybersecurity of automobiles with various organizations on a national level.

South Korea launched the world's first 5G commercial network at the PyeongChang Winter Olympic Games in February 2018. In 2019, South Korea became the world's first country to launch 5G urban autonomous vehicle driving testbed comprising recharging station for electric vehicles, detailed 3D maps, and future mobility center at Sangam in western Seoul. Since then, 5GAA had jointly organized a workshop with 5G forum and ITS Korea in 2019. The 5G forum intends to be the leader of next-generation communications technology and promote the economic growth through the development of the ICT sector in attempts to update the creative economy's new administration agenda. The ITS Korea supports mutual cooperation between the public and private sectors with a view to successfully deploy ITS and contributes through research, policy consultation, and technology promotion activities for the growth of the ITS sector.

UK released guidelines for industry and vehicles, the connected and autonomous vehicle (CAV) cyber security principles considering the security of the whole life cycle of the vehicle and its ecosystem in 2017. In December 2018, the British Standards Institution (BSI) released a new policy on cyber security for vehicles based on the CAV cybersecurity principles. The BSI also published two Publicly Available Specifications (PASs). The first is PAS 188532, which is high-level guidelines to ensure cybersecurity, also known as "The fundamental principles of automotive cyber security." The second is PAS 1128133 that provides recommendations for managing safety risks in a related vehicle ecosystem. It is also known as "Connected automotive ecosystems–impact of security and safety-code of practice" [1].

B. *The vehicular communication standardization in Asia and other parts of the world*

As the ITS develops rapidly, regulatory bodies and government agencies all over the world are paying attention to the technology and spectrum issues related to ITS. In other parts of the globe, efforts have been made to harmonize the ITS arrangements with wider global developments so that the vehicle drivers can enjoy the advantages of the V2X technology as they become available.

In 2016, Korea allocated the 5855-5925 MHz band in 5.9 GHz frequency for C-ITS (V2V and V2I communications) based on a 10 MHz channelization by the Ministry of Science and ICT (MSIT) that is similar to EU and USA [2]. The Korean government regulations assigned 5855–5925 MHz band for ITS applications without specifying any specific radio technology to be performed in this spectrum so that any radio technology can be used in this spectrum as long as technology abides with the corresponding regulations. There is an ongoing amendment to the Korean ITS standards in the Telecommunications Technology Association (TTA), and one of its objectives is to support various radio technologies for ITS applications, including C-V2X along with IEEE 802.11p.

In 2015, China introduced Intelligent Connected Vehicles (ICV) project as a national development program to promote made in China 2025. In October 2018, China allocated 20 MHz band (i.e. 5905-5925 MHz) in the 5.9 GHz spectrum to be used as Internet of Vehicles (IoV) accommodating LTE-V2X technology. Until now, China conducted various successful tests using the 20 MHz band on domestic ground.

In 1994, Japan already reserved its radio spectrum for ITS applications in 760 MHz and 5.8 GHz frequency bands based on regulation policies [3]. At the present time, C-ITS for the implementation in one 10-MHz channel at 760 MHz for ITS safety is considered. Japan uses two different spectra for ITS applications; they are 755.5-764.5 MHz band and 5770-5850 MHz band as shown in Table 8.1. However, the Japanese 760 MHz band for ITS overlaps with LTE mobile network in New Zealand.

In 2014, the Land Transport Authority (LTA) and Intelligent Transport Society of Singapore announced ITS strategic vision for sustainable and future urban transportation ecosystem called Smart Mobility 2030. In October 2017, the Telecommunications Standards Advisory Committee (TSAC) proposed a specification for ITS to improve traffic management, security and mobility of transportation, and V2V and V2I communication architecture. It allocated frequency spectrum from 5855 MHz to 5925 MHz, and the spectrum is divided into 10 MHz channel in 5.9 GHz for DSRC operation. In Singapore, vehicle OBUs do not require license to operate, while localized radio communication license or wide area private network license is required to install RSUs or stationary equipment.

The Australian Communications and Media Authority (ACMA), which is a communication and media regulatory body, is currently reviewing the European C-ITS specification ETSI EN 302 571 as a regulatory basis for operations of C-ITS at 5855-5925 MHz in Australia. The ACMA introduced the Radiocommunications

Class License 2017 for ITS, which will support the use of complying wireless technologies and devices. The ACMA does this by means of specific laws, rules, guidelines, and codes of conduct. The regulations make it possible to use the 5.9 GHz band in Australia for ITS and comply with the ITS provisions in major automotive industries including the USA and EU.

C. *EU and International Cybersecurity Policy*

(i) **EU policy**: Since 2004, the European Union Agency for Cybersecurity (ENISA) has been working to secure Europe from cybersecurity issues. ENISA is working with the EU, its member states, private industries, and the EU citizens to develop good practice and advice on information security. ENISA aims to improve current EU member expertise by supporting the development of cross-border communities focused on improving network and information security across the Europe. It has been running certification schemes for cyber security since 2019. In 2014, the Commission's Directorate-General for Mobility and Transport (DG MOVE) established a C-ITS deployment platform. The EU Commission along with C-ITS stakeholders cooperate with each other to define interoperability. They come to an agreement on how to ensure cross-border and value chain interoperability of C-ITS. In 2017, small and medium-sized enterprises (SMEs) sector began an initiative on general and pedestrian safety regulations, to further decrease the number of road fatalities and injuries. In 2018, the Directorate-General for Communications Networks, Content and Technology (DG-CONNECT) initiated a Cooperative, Connected and Automated Mobility (CCAM) to provide additional specification on the governance framework for accessing and exchanging connected vehicle data and give clear view on the cybersecurity requirements for the connected vehicle environment. In 2019, the European Commission set up an informal group of hundred experts to provide advice and support on tests and pre-deployment activities for CCAM. In 2018, the EU General Data Protection Regulation (GDPR) addresses privacy protection of road users and personal information that formally came into operation. The European Union's Network and Information Security (NIS) directive examines the cyber security concerns of autonomous vehicles as it aims to provide uniform security measures to improve the vehicle cybersecurity.

(ii) **International Policy**: Outside of EU, the countries around the globe are establishing and deploying the cybersecurity policy for the automotive vehicles. In the USA, the NHTSA released a document for improving the cybersecurity and best practices for smart vehicles in 2016. This document was intended to cover safety problems for all motorized vehicles and is relevant to all the automotive systems and software manufacturing and developing individuals and organizations such as automotive vehicle manufacturers, suppliers, and designers. Since 2016, the US Automotive Information Sharing and Analysis Center (Auto-ISAC) has retained a series of best practices in automotive cybersecurity that issue guidance on the implementation of the cyber security of the vehicles. In 2018, UNECE drafted a proposal for a cyber-security recommendation focusing on key cyber threats and vulnerabilities against vehicles that need to be considered to alleviate detected cyber-attacks. The UNECE also implemented the UN cyber-security policy that carries out a set of requirements, which will cover the vehicle life cycle (i.e. from vehicle development to its scrapping) for vehicle producers, suppliers, and service providers.

References

1. G Automotive Association, "The Case for Cellular V2X for Safety and Cooperative Driving," *5GAA Whitepaper*, pp. 1–8, 2016
2. TechVision Group of Frost & Sullivan, "Vehicle-to-Everything Technologies for Connected Cars: DSRC and Cellular Technologies Drive Opportunities.," 2017
3. GAA, "Timeline for deployment of C-V2X – Update," 2019